THE CLASSIC CONNECTION

Peter Pryor

THE CLASSIC CONNECTION

PETER PRYOR

Foreword by Bill Curling

CORTNEY PUBLICATIONS, LUTON

Published in 1979

by

Cortney Publications, Luton
95–115 Windmill Road, Luton, Beds LU1 3XS

ISBN 0 904378 05 5

Printed in Great Britain by
Inprint of Luton (Designers & Printers) Limited
95–115 Windmill Road, Luton LU1 3XS

CONTENTS

FOREWORD

by Bill Curling

Peter Pryor, the author of this book, has been a racing and breeding enthusiast since Oxford University days half a century ago. An owner-breeder on a small scale, a Hertfordshire farmer, a lover of Surtees and Trollope, he has seen most of the big races in Europe since the end of the 2nd World War, and, as his friends know well, has spent many hours doing research into the breeding of the top class horses of the last hundred years.

The result, in effect, is an admirable historical directory of the breeding of the winners of the big European races since 1890, based on the story of fifty one mares of Pryor's choosing and their descendants, and details of the main surviving classic lines:— the dynasties of Bend Or, St. Simon, Isonomy, Hampton, West Australian and the French Gardefeu, whose chief representative in England now, via Tourbillon, is the Derby winner Blakeney.

This is a book for the thorough-breeder. Peter Pryor's researches point to the fact that the majority of the top class horses of this century are inbred in the third or fourth generation, and, therefore, that inbreeding should in the long run pay. His appendix of those closely inbred, those inbred in the fourth generation, and those with no inbreeding for four generations, will be of particular interest.

A study of Pryor's chosen broodmares shows the great uncertainty of the Turf, of which he is well aware. To give four examples: the mare Stella was modestly bred and never ran and yet from her descend the classic winners Trigo, Tulyar and Bruni; Black Cherry ran only once, winning a seller, and yet from her descend the great sire Blandford and the classic winners Santa Claus, Sun Chariot and Carrozza; Eryholme never ran, her dam never won, and from her descend Pharos and Fairway, whilst Sunshot won one small race and from her descend the Derby winner Windsor Lad, Wolver Hollow and Never Bend, the sire of the Derby winner Mill Reef.

This is a book to browse over on a winter evening. Thorough-breeders will assuredly be grateful to the author for the skilful research which has produced it.

Bill Curling, formerly Racing Correspondent
of The Daily Telegraph

INTRODUCTION

The methods employed in the production of the British Thoroughbred, thanks to research and the possibilities of the computer, are now turning in the direction of more scientific selection than has been achieved in the past. A great deal is being done to improve the methods of our ancestors but it is anybody's guess how long it is going to take, either to improve the performance of winners of great races or to upgrade – and that must mean to make go faster over a reasonable distance – the breed as a whole.

The procedures of the past are easily held to ridicule. Whatever guided breeders during the last hundred years – Dosage systems, Family Numbers, Lord Wavertree's astrology (he bred the winners of 4 classics), convenience, conformation or mere accident – this book is a study of what happened to create the top animals in European racing since the early 1880s. If they were the results of methods, then the methods worked; if hit or miss, then these were the hits.

The facts which I retail are discoverable in any books of reference, particularly the General Stud Book and Bobinski's wonderful Family Tables on which I have drawn gratefully. However, my object is to present these facts in a rather romantic form, through the equine heroines and heroes which have appeared since St. Simon (1881 to 1908). He played such a tremendous part in the pedigrees of so many horses (he bred the winners of 571 races and over £550,000, no mean sum at the turn of the century) that even if it is extravagent to claim that he started a new era, at least he makes a good point at which to open.

The approach is from 51 mares and 6 dynastic lines of stallions; the book is in fact a directory, with brief descriptions of breeding and performance of every great horse connected with Europe's biggest races since 1881. While not specifically dealing with all these winners, I have concerned myself with the animals which appear to have contributed most to the breeding of the winners of the following 14 races:–

- All 5 British 3yo classics
- The French Oaks, Derby and Grand Prix de Paris
- The Ascot Gold Cup (nowadays considered an almost lethal ingredient of any pedigree but which played a large part early on)+
- The Irish Sweeps Derby – since 1962
- The Eclipse Stakes
- The King George VI & Queen Elizabeth Stakes
- The Prix de l'Arc de Triomphe
- The Champion Stakes

+Since writing this the British National stud has acquired the 6yo SAGARO (Appx. B), the only horse ever to have won the Gold Cup 3 years running.

I shall try to show some of the aptitudes and characteristics which, over 4 generations, went into the genetic make-up of these extraordinary horses, for that is what they were when one considers the things that can and do go wrong to prevent (a) the breeding of a good foal (b) its ever seeing a racecourse and (c) its winning even the tiniest race. I repeat, I confine this study to some of those concerned in the winning of the 14 greatest races in Europe.

(Throughout the text I use the word "classic" when referring to these races. I realise that this is wrong, there being strictly only 5 classics in all countries, and they for 3yos only; I crave indulgence, for it seems to me that the invention of a new word would be more confusing).

The first part concerns 51 mares, not, as such, the great race winners – they cannot all be Pretty Pollys or Sceptres – but I review the origins of those which have become matriarchs in that their daughters have founded female blood-lines leading to classic winners and to the sires which have contributed to them. For reference only, I append Bruce-Lowe Family numbers. Nor are these mares always the dams of big winners; for instance I exclude Perdita II who presented St. Simon with the incredible brothers Florizel II, the all-conquering Persimmon and the triple crown winner, Diamond Jubilee; I omit her sadly for the simple reason that her daughters have not yet led to anything that has concerned the classics.

It is necessary to face the fact that in the realm of thoroughbred breeding, successes always have been in the minority (I refer to racing as opposed to the Sale Ring) compared to failures and near-misses; also climate, soil, handling, mishandling, accidents and such environmental phenomena have all too frequently been the deciding factors.

If there is romance in this book, it is based nevertheless on the assumption that since each parent passes on one half of its genes, it is uncertain, though far from impossible, that aptitudes and characteristics are transmitted and traceable over more than 4 generations. I therefore make no note of inbreeding beyond the 4th remove.

I then trace, in the second part, the dynasties of the classic sires in a repetitive way which I hope makes clear the intricacies of their breeding which have assisted in the survival of these tail-male lines.

The incidence of great mares in their make-up is particularly interesting. There are 92 of these stallions and they include the winners, since the turn of the century, of 74 classics. Nearly a third of them descend from the mares in part one.

The pages which follow advance no theories and will show nobody how to breed winners, but I hope that such patterns as are discoverable in the matings – e.g. stayers with sprinters etc. – and the incidences of inbreeding and outcrossing will be as interesting to the reader as they are to me. Ap. D. (inbreeding) seems to point to the fact that, in the past, inbreeding has paid as regards the very top class.

Next follows a list of Abbreviations and Signs which I am producing in duplicate so that one copy can be used to avoid referring back.

ABBREVIATIONS & SIGNS

	The winner of the	**miles**		
2	Two Thousand Guineas	1	3yos	
1	One Thousand Guineas	1	3yos	fillies
D	Epsom Derby	1½	3yos	
O	Epsom Oaks	1½	3yos	fillies
FD	Prix du Jockey Club (French Derby)	1½	3yos	colts
FO	Prix Diane (French Oaks)	1¼	3yos	fillies
GC	Ascot Gold Cup	2½		
GP	Grand Prix de Paris	1,7½f	3yos	
IRD	Irish Sweeps Derby (since 1962)	1½	3yos	
GVI	King George & Queen Elizabeth Stakes	1½		
E	Eclipse Stakes	1¼		
L	St. Leger	nearly 1,7f	3yos	
AT	Prix de l'Arc de Triomphe	1½		
Ch.	Champion Stakes	1¼		
=	Dead-heat			

Subsidiary races & local classics on which the text is NOT based but which are mentioned when applicable:–

(F1)	Poule d'Essai des Pouliches (French 1000 gns)	1	3yos	fillies
(F2)	Poule d'Essai des Poulains (French 2000 gns)	1	3yos	colts
(Ir.1)	Irish 1000	1	3yos	fillies
(Ir.2)	Irish 2000	1	3yos	colts
(Ir.D)	Irish Derby (up to 1962)	1½	3yos	colts
(Lup)	Prix Lupin	10½f	3yos	in May
(Hoc)	Prix Hocquart	1½	3yos	in May
(FL)	Prix Royal Oak (French Leger)	1,7f	3yos	
(Wash.Int.)	Washington International, USA	1½		
(Bel)	Belmont Stakes, USA	1½		
(KD)	Kentucky Derby, USA	9½f		
(Preak)	Preakness Stakes, USA	9f		

and the following Stakes:– Blue Riband Trial (3yo 8½f), Jockey Club Stakes (1½m), Chester Vase (3yo 1½m), Coronation Cup (1½m), St. James' Palace (3yo 1m), Hardwicke (1½) and these Cups – Goodwood (2m5f), Doncaster (2m2f) and Jockey Club (2m). Also these handicaps are sometimes mentioned:– City & Suburban (1¼m), Jubilee (1¼m), Cambridgeshire (9f) and Cesarewitch (2m2f).

A * before a sire's name means he sired two classic winners and/or was one of the top ten *sires of winners* in his day.
A † before a sire's name means he sired the dams of two classic winners and/or was one of the top ten *sires of broodmares* in his day.

These symbols will probably take some getting used to and may well be annoying, but they play a very important part in what follows by saving the repetition of many much more annoying eulogistic adjectives.

In the sires' pedigrees an *ANIMAL* underlined shows inbreeding.

In the notes lack of space has driven me to such abbreviations as e.g. "KSAR, inbred Omnium II 2 & 3 rem." which means that Omnium II appears once in Ksar's 2nd remove & once again in his 3rd.

The story of POCOHONTAS

Before chronicling the 51 mares I must tell the story of one from the past who features, sometimes more than once, in the backgrounds of most of the mares I record up to 1914;

POCAHONTAS 1837, by †**Glencoe** 2 GC ex **Marpessa** by Muley

She was a roarer and a moderate racehorse; this is her record at stud –

year of foal	had been mated with	Product
1843	Camel	c Cambaulcs, a roarer, never won
1845	Camel	c unnamed, never ran
1846	Muley Moloch	f Dolly Varden, never won
1848	Muley Moloch	f Indiana, never won, remote ancestress of winners of 8 classics including Kizil Kourgan (see later).
1849	The Baron	*†STOCKWELL 2 L, won 12 races, 7 times champion sire
1850	The Baron	*†RATAPLAN, Stockwell's brother, won 42 out of 72 races inc. the Doncaster Cup, sired the Grand National winner Austerlitz.
1851	Harkaway	*†KING TOM, won twice, 3 times champion sire, sire of St. Simon's dam.
1852	Chatham	c Strood, never won.

In which year Stockwell won his two classics and Pocahontas would have been recognised now as having maternal merit. Meanwhile Stockwell's & Rataplan's sire had been sent to France.

1854	Birdcatcher	f AYACANORA, won twice, dam of Chattanooga (sire of Wellingtonia – see below) & ancestress of classic horses and of 3 notable stallions in USA.
1855	Nutwith	c Knight Of Kars, won twice; among some other minor successes at stud, he sired, from a ½bred mare, The Colonel, twice winner of the Grand National.
1856	Nutwith	f Heroine Of Lucknow, never ran, 4th dam of Clover FD (by Wellingtonia who was not only her nephew's son but also her nephew – see below and above).
1858	Knight Of St. George	c Knight of St. Patrick, won 4 races, sired Moslem 2.
1859	Ambrose	c Automaton, won twice and died aged 4.
1860	Ambrose	f AURICULA, won 3 races, dam of the stallion Blandford 1873 (not the famous one).
1862	Ambrose	f ARAUCARIA, never ran, dam, by her nephew Chattanooga, of Wellingtonia (see above) & dam also of the winners of 4½ classics.

She died in 1870 having had 15 foals in all.

Looking at her pedigree, her classic sire Glencoe went back to Herod in 5 removes, his bottom line being crammed with dams of big winners. Pocahontas was the first foal of a dam of promising ancestry, in fact an expensively bred mare in her day, with a classic background. Her dam, Marpessa, was not mated again with Glencoe as he was sent to America in 1836.

The pedigrees which follow are full of her son Stockwell who lived to be 21 and sired 17 classic winners. His brother Rataplan occurs quite often and of course Pocahontas's other great stallion son, King Tom, persists in his role of maternal grandsire of St. Simon. Finally I take leave to quote from John Osborne's "Horse-breeder's Handbook 1898" in which the author claims to have seen Pocahontas and

'thought her lacking in quality, but was much struck with her fine size and great bone, which she happily transmitted to most of her progeny, but particularly to Stockwell, Rataplan and King Tom, through whom, as well as by her daughters, her blood is to be found in nearly all the best runners of the present time.'

Part One THE MARES

ALL DATES, unless otherwise specified, are DATES OF FOALING

The key, if there be one, to the success of these female lines, must be, supplementary to their own female merits, the sires to which they are mated – the new blood introduced along the generations. Sires are entered in small letters and a description of most of them can be found through the index and/or Appendix A.

CONNIE 1884

The origins of **CONNIE 1884** by the stayer Pero Gomez 1866 L.

Ex HILARITY 1871	by the stayer *†King Tom (p.188). Ran 31 times, winning 7. Had 14 foals, 11 of them by Pero Gomez. Of these 8 sisters & 3 brothers Connie, who won once at 2y over 10f, was the only one to achieve anything. I have often wished that breeders, having sat up all night planning their matings, could stick to the same stallion on more occasions instead of switching, before the result can be known, to some other sire. However, despite Hilarity's relentless fidelity to Pero Gomez having produced 90% duds, too many success-stories of brothers and sisters seem to occur throughout history to make me change my mind.
Ex NIGHTINGALE	by The Mountain Deer, dam also of *Uncas.
Ex CLARINDA	by Sir Hercules, 3rd d. of the 1881 Grand National winner Woodbrook.
Ex MUSTARD	by Philip The First, a grand-daughter of the key mare WIRE 1811. Family no.1.

This is a fairly strong pedigree with 2 strains of **TOUCHSTONE** at the 4th remove. At the time of mating, Pero Gomez had already produced a winner of the 2000 gns. and King Tom by that time was famous as a classic sire.

Connie bred 7 living foals in all, 3 of them by sires of no consequence. It is worth noting the high class sires which have contributed towards keeping her descendants up to classic level – 21 of them (19 different), the winners between them of 21 classic races.

CONNIE'S MATINGS:

MERRY HAMPTON 1884 D – not a particularly successful stallion, his only other product of interest being the stayer Pride 1892 who appears in the far background of High Hat's pedigree.

*†**JUGGLER** 1885 – a sprinter and a good sire of mares; as well as Conjure his daughters bred Perola O & the stallions Pilliewinkie & Torloisk.

CONNIE 1884

Mated with: –

Produce: –

MERRY HAMPTON

1891 MERRY WIFE
did not win, bred
SANTOI 1897 GC
Queen's Birthday
At stud first in Ireland
then exp. Germany.
Sired †Santorb GC,
Shogun, *Achtoi and
He.

JUGGLER

1895 CONJURE
did not win, bred 5
good wnrs. including
the sisters

THIRD TRICK
1906 William The
Third, won 2 races,
bred 4 wnrs.

and

WINKIPOP 1907 1.
William The Third
(exp USA '27)

PLYMSTOCK 1918
Polymelus
won 3 races, bred
10 wnrs. £31,000

PENNYCOMEQUICK
1926 O. Hurry On

PICTURE 1925
Gainsborough
won 2 small races

INSTANTANEOUS
1931 Hurry On.
Won 1½ races

***† COURT MARTIAL**
1942 2, Ch. Fair Trial
exp. USA '58 Inbred
Polymelus 4 rem. Won
also 4 other races, max
10f. Sired Timandra F0,
*Counsel, †King's Bench,
†Ratification, †Above
Suspicion and the dams
of Martial 2, Only for
Life 2, and the luckless
Rock Roi (disq. twice GC)

PIN PRICK
1917 Torpoint
won once, bred
7 wnrs.

POINT DUTY 1926
Grand Parade
won 3 good races,
bred 5 wnrs.

LONG SUIT
1918 Lemberg
won once, bred
3 wnrs.

SHORT STORY
1923 O. Buchan
inbred Cyllene
at 3 and 4 rems.

TRAFFIC LIGHT
1933 Solario
won 3 good races
bred 9 wnrs.

AMBER FLASH
1942 Precipitation
won 5 races, bred
4 wnrs.

AMBIGUITY 1950
O. Big Game
exp USA '66

and her sister

GAMBADE
1951 Big Game
did not win

SODIUM 1963
IRD L Psidium
exp. FR. 1972

THORNWOOD
1942 Bois Roussel
won twice, bred 7
wnrs.

THICKET 1947
Nasrullah, won once
bred 5 wnrs.

***† HORNBEAM**
1953 Hyperion
raced 4 seasons,
won 10 races up
to 2 miles. Sired
Intermezzo L and
Windmill Girl, d.
of Blakeney D
and Morston D.
Exp. Sweden 1966.

FAIR RANGER
1943 Bois Roussel

SUNSET III 1952
Hyperion exp. USA
1953

MOON MADNESS
1963 Tom Fool

MADELIA, 1974
(F1) F0 Caro

RINOVATA 1887

The origins of **RINOVATA 1887** by the stayer *Wenlock 1869 L who, at the time of this mating, had probably not achieved much fame at stud, though later daughters of his were to produce Isinglass and Sainfoin.

Ex TRAVIATA 1881	by the stayer *Cremorne D GP GC, did not win exp. France 1887
Ex THE WHITE LADY	by King Of Scots, exp. Fr. '86
Ex BLANCHETTE	by Destournel
Ex LADY BLANCHE 1858	by *†Voltigeur D, L. Family No. 2. Her other daughter, May Queen, bred 3 classic ancestresses of Rosedrop O, *†Gainsborough 2 D L GC, *†Teddy FD & Dhaudevi GP.

Rinovata won one race; she was foaled in UK and inbred to Rataplan at the 3rd & 4th removes and her pedigree contains 3 strains of Pocahontas, King of Scots having been sired by Rataplan's half brother King Tom. Rinovata had 9 living foals in all, the others were of no consequence.

RINOVATA'S MATINGS:

***DONOVAN** 1886 D, L & 16 other wins; he missed the triple crown by a head in the 2000. A disappointment on the whole as a sire although he got *Matchmaker and the dams of 3 classic winners. By Galopin ex **Mowerina** 1876 by The Scottish Chief.

*†**ST. SERF** 1887 by St. Simon from the good mare Feronia by Thormanby. A middle-distance horse of no great merit on the racecourse, like not a few of St. Simon's sons, he became a first-class sire getting 2 classic winners, the dams of 3 and the stallion Ormondale.

A note about *†**ORBY** 1904 D (Ir.D) who appears 4 times himself (next page) and again as the sire of Grand Parade & Flying Orb. By the 10f champion *†Orme E^2, Ch. out of the American mare **Rhoda B** imp. UK 1896 by the stayer Hanover, he was himself intended as a stayer, but broke down in training for the Leger. However, despite having got **Grand Parade** D and **Orpheus** $Ch.^2$, the average winning distance of his progeny was only 6. In fact the presence of Orby in a pedigree indicates SPEED and that, though its influence was of untold value to the breed, knocked him out of the direct line of classic progenitors. It will be seen how the infusion later of staying blood from Son In Law, Gainsborough etc. has kept this family's winning distances up to classic level.

RINOVATA 1887

Mated with: – DONOVAN

Produce: – 1900 DONETTA ran 47 times, won 13 (6 to 10f). Bred 6 wnrs. of 43 races £47,800

- DONA SOL 1908 Ayrshire did not win
  - LADY PHOEBE 1917 *Orby* won twice, bred 5 wnrs.
    - LADY DISDAIN 1924 Hurry On did not win
      - COLOUR BAR 1930 Colorado did not win exp. France 1936
        - COULEUR **1939** Biribi (Appx. A) won once (hurdles)
          - **CAMAREE 1947 1. Maurepas** & her sis.
          - MARMALADE 1949 Maurepas
            - SICALADE 1956 Sicambre
              - ***SEA BIRD II 1962 D,AT,(lup.) Dan Cupid** won also 4 other big French races, was beaten only once, in the Grand Criterium in the mud, when he finished fast to be second. Exp. from France to USA & back to France 1973 where he died. Sired Allez France (F1) F0, AT. (see ap. D)
    - LULA 1925 Capt.Cuttle did not win
      - LUCIEBELLA 1937 Rodosto did not win bred 4 wnrs.
        - **LUMINARY 1946 Fair Trial** exp. USA '59 av. winning distance of 6.5f sire of d. of Canisbay 1961 E.
- **DIADUMENOS** & sis. **1910** *Orby* sprinter exp. Germany 1923 sire of d. of *†Gold Bridge
- **DIADEM** 1914 *Orby* **1.**
- ***†DIOPHON** (exp.Greece '35) **1921 2. Grand Parade** Sired Diolite 2 and the d. of Turkhan L and Ujiji GC

Mated with: – ST. SERF

Produce: – 1902 RENAISSANCE won 2 good races bred 8 wnrs. of 26 races £16,900

- EOS 1913 *Orby* won 6 races
  - ARS DIVINA 1925 Gainsborough won 1 small race
    - BEAUTIFUL GIRL 1932 Son In Law did not win
      - COSMETIC 1940 Sir Cosmo won once, bred 8 wnrs.
        - **ROCKAVON 1958 2. Rockefella** inbred Gainsborough 3 & 4 rems. Rockavon was a sub-classic sire; exp. France 1970. His sire, *ROCKEFELLA 1931 won 3 small races, (av. 13.3f) was by Hyperion from the Felstead mare ROCKFEL 1, O, Ch. who bred nothing else of note. Sired also ANGELA RUCELLAI (It. 1, 0) d. of *Appiani II (It. D).
- COS 1920 Flying Orb (exp.Fr '37) won 8 good races, exceptionally fast
  - **†RUSTOM PASHA 1927 E, Ch. Son In Law** Stood first in France ('37) then in Argentina: sired the mare Rustom Mahal (p.126 App.D) and the d. of Lavandin D and Le Lavandou.

ORNAMENT 1887

The origins of **ORNAMENT** 1887 by ***BEND OR** 1877 D, Ch. – when this mating took place (1886) he would not have been famous as the sire of the triple crown Winner **Ormonde, sis. to Ornament**.

Ex LILY AGNES 1871	by *†MACARONI 1860 2, D. Ran 32 times, won 21 races up to 3 miles. d. also of Farewell 1.
Ex POLLY AGNES	by The Cure, gd.dam of the stayer *†Martagon; her 5 daughters (4 of them sisters to Lily) led to 2 other classic families.
Ex MISS AGNES	by Birdcatcher; her daughters & gd. daus. founded 5 classic lines between them.
Ex the founder mare AGNES 1844	by Clarion, did not win.

The AGNES ramification is responsible in all for about 50 classic wins up to the Oaks of 1969, with Petition & Crepello thrown in among many other important stallions. Family no. 16.

Ornament had 11 foals to live.

ORNAMENT'S MATINGS:

***†HAMPTON** 1872, a stayer, won Goodwood and Doncaster cups and the Great Metropolitan h'cap at 3y; he also ran over hurdles at that age. More of him later as the forbear of Hyperion.

***SHEEN** 1885, by Hampton, also a stayer who won the Cesarewitch and Jockey Club Stakes.

BLUE GREEN 1887, won from 8f to 3 miles, by the not very distinguished Coeruleus out of St. Simon's sister Angelica (see later). Brooch, thanks to him, has Stockwell twice at the 4th remove.

***†PERSIMMON** 1893 D, L, GC, E, by St. Simon from the Hampton mare Perdita II, was the middle one of the 3 famous brothers (Florizel II & Diamond Jubilee being the other two). He became 4 times champion sire of winners and 3 times of brood mares and might well have done greater things had he not died at the age of 15. As it is, in addition to Sceptre, his offspring accounted for 3 Eclipses, 3 Gold Cups, 2 Oaks and a Leger.

ORNAMENT 1887

Mated with:–	HAMPTON	SHEEN	BLUE GREEN	SHEEN	PERSIMMON
Produce:–	1892 STAR RUBY first foal exp. USA 1894, did not win in UK. Reappears in pedigree of Flossy 1966 Ch.	**1893 LABRADOR** **Ch. exp. Belgium**	1894 BROOCH won twice at 2y bred **1901. ST. DENIS** **St. Simon** useful racehorse 3rd in D. Inbred Galopin 3 & 4 rems.	Labrador's sister 1896 SPLENDID did not win bred FINE FEATHERS 1904 Gallinule exp. France 1917 TUSCAN RED 1912 William Rufus exp. USA 1917 WAR FEATHERS 1924 Man O' War MISS KIDD 1930 Atwell BURGOO MAID 1940 Burgoo King SECRET MEETING 1950 Alibhai SANTORIN 1956 Greek Song **LONG LOOK** **1962 O** **Ribot** inbred Hyperion 4th rem.	**1899 SCEPTRE** **2, 1, O, L** **founded a family of her own.** **(p. 65)**

CHARM 1888

The origins of **CHARM 1888** by the great stayer and greater sire, ST. SIMON GC. See p. 23, 161, 168, App. D, who was only 6 when Charm was planned. She won twice.

Ex TACT 1882	by the useless racehorse *†WISDOM 1873. His sire was a long range stayer, Blinkhoolie, and his two grandfathers were the brothers Rataplan & Stockwell which made him inbred at the 3rd rem. to The Baron & Pocahontas. He never won and when mated with Enigma had achieved nothing at stud. The causes of his having become a very successful sire later on give rise to some speculation. Tact did not win; d. of Charm's sis. Aimiable 1891 1,0.
Ex ENIGMA 1872	by The Rake.
Ex THE SPHYNX 1865	by Newminster. Won 6 sprints and founded 2 classic families.
Ex MADAME STODARE	by Sleight of Hand. Family no. 2.

CHARM had 13 foals to live, all of them to stallions of some renown. Only Fascination and Captivation appear to have been of much use, though Fascination's brother, Red Robe, won the Goodwood Cup.

CHARM'S MATINGS:

***ROYAL HAMPTON** 1882 a moderate middle-distance handicapper. By Hampton from a mare by King Tom, which means that Fascination is inbred to the latter at the 3rd & 4th removes. He sired Marcion GC, Kirkconnel 2, †Phoenix, Forfarshire & the dam of †By George.

*†**CYLLENE** 1895 GC (at 4y), did not run in the 3yo classics, being thought too small at the time. He was impeccably bred by Bona Vista 2 from an Isonomy mare. He became champion sire of winners twice and of brood-mares once, getting 9½ classic winners the greatest of which was the next progenitor in the Bend Or line, *†POLYMELUS Ch. Cyllene was exported to Argentina in 1908, it was perhaps a pity that Charm didn't visit him again, but certainly Captivation gave no encouragement until too late.

CHARM 1888

The earliest of the 4 St. Simon mares in this book

Mated with: – ROYAL HAMPTON

Produce: – 1896 FASCINATION
won Coronation Stakes
and 3 other good races

bred

PERFECT PEACH
(exp. USA '25)
1907 Persimmon
did not win
bred 4 wnrs.

†STEFAN THE GREAT
1916 The Tetrarch
Inbred St. Simon 3 & 4 rems. and to Hampton at 4th.
A very fast 2yo, broke down at 3.
Sired Tolgus and the dams of *†BLUE PETER 2, D, E, †PORTLAW & PEACE CHANCE (Belmont). Exp. USA 1923, returned UK later.

FARMAN 1909
Velocity
Inbred Hampton at 3rd rem. Mat. gd-sire of †Finglas

PICARDEL
1915 Picton
bred 8 wnrs. of 31 races
£20,000.

HERBALIST
1923 Phalaris
exp. Fr '27.
Inbred Hampton & St. Simon 4 rem. won up to 7f sired 3rd d. of Armistice 1959 GP.

DEVONSHIRE HOUSE 1924
Swynford
did not win

LOUISE 1932
Finglas
won 1 good race

EDVINA
1940 Figaro
won 7 small races.

FAIR EDWINA
1946 Fair Trial
won twice

FAIR ALYCIA
1952 Alycidon
did not win

BARN PRIDE
1957 Democratic

***BOLD LAD (Ire)**
1964 Bold Ruler
champion 2yo, won at 7f at 3y. His sire *BOLD RULER 1954 (Preak.) was by Nasrullah ex Miss Disco by Discovery. Bold Lad has sired WATERLOO 1969 1.

DISCIPLINER 1948
Court Martial
did not win bred 8 wnrs.

MARTIAL 1957 2.
Hill Gail (ApB. imp. UK '54) exp. Argentina 1967.

FESTUCA 1929
Phalaris
did not win

HAYMAKER
1939 Hyperion
d. of 5 wnrs.

HOT SPELL 1944
Umidwar
won 5 small races

***FIRESTREAK**
1956 Pardal
Inbred Phalaris at 4th rem.
High class mid. dist. h'cpr. won 9 races
Sired Snow Knight 1971 D.

Mated with: – CYLLENE

Produce: – **1902 *CAPTIVATION**
ran only once and was not placed. Began life at stud as an Irish country sire at £5 a time and very few good mares were sent to him; however, he got a lot of winners, though only one of any consequence – KIRCUBBIN 1918 who, having won the Irish Leger, went to France in 1921 and sired the French Derby winner Chateau Bouscaut thus becoming the progenitor of *Pinza D, GVI. Captivation's daughters played leading parts in the pedigrees of Exhibitionist 1,0 *†Pharis FD, GP, Arctic Star, Arctic Prince D, Peter Flower Ch., *†Solar Slipper and Tropique E.

DAME MASHAM 1889

The origins of **DAME MASHAM** 1889, who never ran. By *GALLIARD 1880 2. exp Ger. '95.

Ex PAULINE 1883 — by *†**HERMIT** 1864 D (winner of the sensational 1867 Derby – snowstorm that day – lame in paddock – huge gamble by owner Chaplin). A blood-vessel breaker who became 7 times champion sire of winners, 5 of brood-mares in UK and then champion in France in 1887. Yet his tail-male descendants have been barely on the edge of the classics for generations and don't look like getting any closer. However, I hope his mark is indelible thanks to his son *†TRISTAN 1878 GC, Ch.3 having sired the mare Canterbury Pilgrim (p.37, 167, 168, App. D), dam of both Chaucer and Swynford. Pauline did not run.

Ex LADY MASHAM 1867 — by Brother to Strafford, bred 3 brothers to Pauline:– †Peter, a good wnr. and sire of Primrose FO, Paul (undistinguished) and Timothy GC, exp. Austria.

Ex MAID OF MASHAM 1845 — by Don John. Ran 35 times winning 23 races, bred 12 foals, of which 4 daughters founded classic families.

Ex MISS LYDIA — by Belshazzar. Family no. 9.

Dame Masham had 7 other foals to live.

DAME MASHAM'S MATINGS:

*†**BEND OR** 1877 D, Ch. Founder of the most powerful dynasty now extant. See p.147, 177.

†**AMPHION** 1886 Ch. & 13 other races inc. the Hardwicke Stakes. By Rosebery (a stayer) from a Hermit mare. Sired the winners of 2 classics and, most important of all, the sprinter *†SUNDRIDGE of whom much more will be heard later. His daughters bred Hapsburg E, Ch., *†Dark Legend, Celt (USA) & Cadet Roussel. The mare Ferelith was inbred to Hermit at the 3rd rem., with 3 strains of Newminster at the 4th and also 2 of Vedette at the 4th.

DAME MASHAM 1889

Important exports to USA and France

Mated with: –

Produce: –

BEND OR

1896 FAIRY GOLD
won 3 races up to 6f
at 3y Exp. USA
1903

bred

1901 ST. LUCRE
exp. France 1908
St. Serf.

GOLDEN LEGEND
1907 Amphion

***†DARK LEGEND**
1914 Dark Ronald
a stayer, won 3 times
UK then several good
races in India, return-
ed 1917 and went
to France where he
sired the winners of
5 classics, Easton,
Legend of France and
the dams of an FO
winner and of *†Dante
1942 D. Returned UK
1920

GOLDEN ORB
1916 Marcovil
a sprinter, inbred to
Speculum and Hermit
4 rem. Sired the d. of
the sprinter Bellacose
(mat.gd-sire of *Hard
Sauce)

ZARIBA 1919
Sardanapale
founded family
of her own p.121-123,
App. D.

***†FAIR PLAY 1905**
Hastings
3 times champion sire
USA (p. 76)

FRIAR ROCK
1913 (Belmont)
Rocksand
Inbred Galopin
& Hermit 4 rem.
Sired Pilate &
the d. of War Relic
(p. 177)

AMPHION

1898 FERELITH
did not run, does
not appear to have
bred a UK wnr.

bred

FERULA 1907
St. Serf
exp. France 1912
did not win

DIANA MALLERY
1917 Nimbus

LIKKA 1925
Sardanapale
won twice

ASTRONOMIE
1932 Asterus
won twice, d. of
9 wnrs.

I regret not being able to give Astronomie a whole page to herself, but Asmena's offspring have yet to appear at classic level.

***MARSYAS II**
1940 Trimdon
wonderful stayer, Alex.
Stks. Don. & Gdwd.
Cups, Cadran 4 times.
Sired Macip GC
His sire TRIMDON
1926 GC^2, by Son In
Law ex Trimestral by
William The Third
(dam of Foxhunter
p. 119)

CARACALLA II
1942 GP, AT, GC,
(FL) Tourbillon
sired d. of *Wolver
Hollow E.

***ARBAR 1944**
GC (Cad.) Djebel
Inbred Teddy 3 &
4 rems. sired d. of
Crepellana FO

ASMENA 1947 O
Goya II
Inbred Sardanapale
3 rem., both her
gd.parents being
by him.

BEND OR

1899 DAME D'OR
Sis. to Fairy Gold
did not win. d. of
6 wnrs. of 26 races

bred

ALL GOLD 1908
Persimmon
won 1 mod. race
exp. USA 1912
inbred Galopin 3 & 4 rems.
sired 4th d. of Traffic
(USA).

MRS. BUTTERWICK 1890 O

The origins of **MRS. BUTTERWICK** 1890 O by St. Simon who was still not proven as a sire.

Ex. MISS MIDDLEWICK 1876	by The Scottish Chief 1861 GC (at 3y) exp. France 1884. Had 11 other foals, all by notable sires. Her sister Violet Melrose was d. of *†Melton D, L (see later).
Ex VIOLET 1864	by *†Thormanby 1857 D, GC. Sister to Feronia the d. of *St. Serf and gd-dam of *†Ayrshire 2, D, E.
Ex WOODBINE 1850	by Stockwell. Her daughters founded 3 classic families.
Ex HONEYSUCKLE	by Touchstone. A daughter of the key mare Beeswing. Family no. 8.

Little Mrs. Butterwick won 5 other races besides the Oaks. One wonders what the odds would be in the 1970s against an Oaks winner whose sire and mat. gd-sire had both won a Gold Cup.

MRS. BUTTERWICK'S MATINGS:

Her first success, Phaleron, was in fact her 6th living foal and one can't help wondering what would have happened had she gone to ***†GALLINULE** 1884 again. He was already proven as a sire having got the 1898 Leger winner Wildfowler and Pretty Polly (of whom of course much more later) and she, by this time had won 4 classics. On the other hand Phaleron's merits, such as they were going to be, were unknown quantities; but what *was* known was that Gallinule, despite his recent stud successes, was unsound and had had a very mixed career on the racecourse. He made a noise and was a blood-vessel breaker and although he won the Imperial Produce Stakes and other races as a 2yo, his infirmities seem to have caught up with him and to have prevented any further successes. He ended up as a 5yo carrying the pathetic weight of 7.9 in the Lincolnshire and breaking a b.v. just when he might have won. He sired the winners of 8 classics and his daughters bred the winners of 7. As will be seen later he was by Isonomy from a very strange mare, Moorhen by Hermit. He was inbred to Stockwell at the 3rd & 4th rems. Whatever may have accounted for it he was certainly a remarkable distributor of female excellence.

***†ISINGLASS** 1890 2, D, L, GC, E, by Isonomy (stayer) out of a Wenlock mare which never ran and had been thrown in gratis when her previous owner bought a hunter. He was to sire the winners of 5 classics and many great mares.

VELOCITY 1902 won 2 Doncaster Cups and the Cambridgeshire; bred by a sprinter called Speed from a mare of no apparent promise, he sired the stallions Wildair and Farman.

LONG TOM 1899 was a stayer by Ladas from a Bend Or mare.

MRS. BUTTERWICK 1890 O.

The second St. Simon Mare

Mated with: –	GALLINULE	ISINGLASS	VELOCITY	LONG TOM
Produce: –	**1906 PHALERON** a stayer, awkward, won Jockey Club Stakes, sired PEARL MAIDEN (p. 119)	1908 ABBAZIA did not run, bred 6 wnrs. | TETRABAZIA 1918 The Tetrarch won 3 good races, bred 4 wnrs. | ***†SINGAPORE** **1927 L** **Gainsborough** (exp. Braz '43) Inbred St. Simon 4 rem. Also won Doncaster Cup; sired Chumleigh L, the dam of *†Ballymoss L, AT, E, GVI (p.155, App. D) and the 2nd d. of *Charlottesville FD, GP.	1909 BUTTERMERE did not run, bred 1 wnr. | 1917 SADDLEMARK Polymelus did not win | PORTREE 1927 Stefan The Great won 2 small races bred 7 wnrs. | **PORTLAW 1928** **Beresford** Inbred Cyllene and Persimmon 4 rem. A very successful sprinter. Sired †Warrior (mat. gd- sire of Birum GP) and the dam of *Pall Mall 1955 2. Also plays an important part in the pedigrees of Caergwrlie 1, Never Too Late 1, O, Ribocco IRD,L and his brother Ribero IRD, L and *Busted E, GVI.	1913 THOMASSINA did not run, exp. France 1920 bred | TAMBOURINE 1921 Clarissimus | **SIRTAM** **1933** **Sir Nigel** Inbred St. Frusquin 4 rem. Sired d. of Bobar II 1958 Ch. and the 3rd d. of Dhaudevi 1965 GP, (FL)

STELLA 1890

The origins of **STELLA** 1890 are far from imposing, by the sprinter and miler ***Necromancer** 1882, a gt-grandson of Stockwell, by the Lincolnshire winner Touchet; he was a roarer and brother to the quite successful *†Juggler.

Ex HOLLYLEAF	by **Hollywood**, whose name appears nowhere else at classic level. Hollyleaf did not run.
Ex MADELINE	by Plum Pudding, bred Buckshot, a winner over 4 miles and gd-sire of the double GC winner The White Knight; she also bred Boulevard who appears far back in Solonaway's pedigree.
Ex TAWNY	by Ivan. Ancestress of Queenpot 1 and of Petingo.
Ex CAPRICE	by Coronation. Family no. 22.

Stella did not run.

She bred 11 other foals, 17 in all, Blakestown was her 6th.

STELLA'S MATINGS:

†**LESTERLIN** 1892, with the exception of his sire, Gallinule, he came from as uninspiring origins as his bride. He had won races of a mile and less and his fee, at the start of these matings was probably 7 gns. I cannot help feeling that the influence of Gallinule was responsible for the up-lifting process started by these 3 sisters and a brother.

BUSHEY PARK 1889 was a long distance stayer of the line of Hampton, he sired also the dam of Charlebelle O.

*†**ORBY** D (Ir.D) see p.5, 148, App. D.

Looking down to the next generation, **Galloping Simon** appears; he never ran and charged 9 gns. for siring the Irish Oaks winner Athgreany; he was strongly bred by the stayer *†Melton D, L from a St. Simon mare.

Next into this very Irish story comes †**Farasi** 1903; one would have thought that an Irish Oaks winner would have been worth more than the 19gns. (and that later reduced to 5) demanded for his services, but no, this son of Desmond (by St. Simon) honoured the transaction by bringing into being Athasi, whose wonderful union with Blandford is noted overleaf. She also bred the dam of **Artist's Son**, sire of the 1955 Grand National winner, Quare Times.

STELLA 1890

Bred 3 sisters whose descendants are still influential

Mated with: –
Produce: –

LESTERLIN

1902 BLAKESTOWN (sis)
won Irish O, bred
4 wnrs.

BENANET 1909
Americus
won 4 races, bred
5 wnrs.

CAMBELL KID
1919 Bridge of Earn
won 6 races, bred
4 wnrs.

GALLANT KID
1926 Sir Galahad
won twice

POTTED MEAT
1937 Sandwich
won 4 tiny races

LINDUS 1944
Dastur
won once

WHIMBREL 1951
The Phoenix
won once, bred
6 wnrs.

WHIMSICAL 1958
Nearula
won once, bred 5
wnrs.

BOMBAZINE
1963
Shantung
won 2½ races

BRUNI 1972 L
Seahawk II
(ap A)

LESTERLIN

1903 FAIRYLAND (sis)
won at 2y, bred
4 wnrs.

ATHGREANY
1910 (Ir.O)
Galloping Simon
bred 5 wnrs.

ATHASI 1917
Farasi
ran until 6yo under
both rules, winning
5 races. Produced
also Harinero and
Primero each the
wnr. of both Irish
Derby and Leger,
Athford Don Cup
and 4 other impor-
tant winners – all
by BLANDFORD

TRIGO 1926
D, L Blandford
not a success at
stud, appears only
as the sire of the
2nd d. of
Margueritte Vernaut
1957 Ch.

HARINA 1933 (sis)
Blandford
won well at 2y

NEOCRACY 1944
Nearco
won 2 good races
at 2y. dam of 9
wnrs.

TULYAR 1949,
D, L, GVI, E Tehran
inbred Phalaris 4
rem. Won also 5
other big races. A
brilliant & sound
horse, to stud
Ireland 1953 & to
U.S.A. 1956 only
to become des-
perately ill. So far
he has shown no
signs of reviving The
Chaucer dynasty.

*SAINT CRESPIN III
1956 E, AT
Aureole
exp. Jap. 1970.
Sire of Altesse
Royale 1, 0
(ap. C)

LESTERLIN

1904 ZENITH (bro)
won 5 small sprints,
bred 2 wnrs.

PERFECTION
1918 Orby
did not win,
bred 4 wnrs. of
£20,700

OPTIMA 1927
Pharos, did not
win

LEGER DAY
1932 Winalot
won 4 races

HIGH PEAK 1942
Hyperion exp.
Australia 1945,
sired *Dhaulagiri

CASTLE GAY
1928 Buchan
bred 4 wnrs.

SUN CASTLE
1938 L
Hyperion
won 2 other war-
time races, died
before going to
stud.

LOANINGDALE
1929 E Colorado
moderate sire,
exp. Uraguay
1941.

LESTERLIN
|
1906 GLENESKY
won 7 sprints, features in the bottom lines of Noor and Taboun 2.

BUSHEY PARK
|
1908 TEMBLEBODEN
did not run, bred 6 wnrs.
|
GLENINA 1911
Galloping Simon
won 2½ tiny races
|
GILFORD 1920
Corcyra
did not win, bred 4 wnrs.
|
BALLYFERIS
1927 Apron
won 4 races at 2y, sired the Irish 1 & O wnr, Sol Speranza, herself the d. of Solferino (Ir. L).
(*†APRON 1920, stayer by Son In Law from a Chaucer mare. Sire also of the d. of Ki Ming 2)

ORBY
|
1911 *FLYING ORB
a sprinter, his daughters bred †Rustom Pasha E, Ch. and the stayer *†Epigram

and his sister

ORBY
|
1912 KNOCKALT
did not win, bred 4 wnrs.
|
CAPTIVATING STELLA
1917 Captivation
exp. France 1922
|
LA GRANDE OURSE
1926 Sans Souci II
|
VOUTE CELESTE
1937 Legatee
|
AURORA BOREALIS
1941 Brantome
|
TROPIQUE
1952 E
Fontenay
also won Coronation Cup. At stud both in UK '56 and France '63, no notable successes so far. Inbred Swynford 4 rem.

PARADOXICAL 1891

The origins of **PARADOXICAL** 1891. By **TIMOTHY** 1884 GC, won up to 3 miles, by Hermit ex Lady Masham p.111 exp. Austria 1897.

Ex INCHBONNY 1883	by *†**Sterling** 1868, a not very distinguished miler who, by this time, would have acquired merit, if not fame, as sire of the GC winner Isonomy 1875 (to become Sterling's successor in the dynasty see p.168), but of more importance at the time, he had already bred the winners of a dead-heated Derby, two 2000s and a Champion Stakes. Inchbonny did not win, sister to Paradox 1882 2, GP, Ch.
Ex CASUISTRY	by The Miner.
Ex LADY CAROLINE	by Orlando.
Ex LADY BLANCHE	by Stockwell.
Ex the famous COBWEB 1821	dam of the brothers *†Bay Middleton 2, D, and Achmet 2, and of Clementina 1. Family no. 1.

Paradoxical did not run, she was exp. to Belgium 1906.

She had another 8 foals to live, mostly by moderate sires. It was *†**MELTON** 1882 D, L, who kept this line up to classic level. His line in tail-male does not survive in exalted circles, however, he got †Best Man (sire of Signorino and of the dam of Straitlace O) and the possibly under-rated Galloping Simon (p. 15); but his influence through the female sex was far-reaching. His origins as regards the top line were more stout than recherche, sired by Master Kildare the son of Lord Ronald, a Stockwell horse which got more hunters than anything else; but his dam's line is superb, his great granddam being the Stockwell mare Woodbine from whom descend many important sires. Thus he was inbred to Stockwell at the 3rd and 4th rems.

Absurdity's first 3 matings were with Sundridge and the third of them produced the good Jest. So *†**SUNDRIDGE** 1898 may now be introduced, bearing in mind the fact that, when Jest was thought of, his fame was not yet established. He started life as a selling-plater never winning over a greater distance than 6f and, although he sired an Oaks and a Derby winner, the average winning distance of his many successful progeny was only 6.7f. He was by the Champion Stakes winner Amphion from a Springfield mare and was inbred to Stockwell at the 4th rem. He stood in France from 1911 to 1919. As will be seen when we come to

(over)

PARADOXICAL 1891

Ancestress of 4 very great Mares

Mated with: – MELTON

Produce: –

- 1901 DOXA (exp. Fr. '19) did not win
  - LADY OF PEDIGREE 1910 St. Denis exp. France 1914
    - HELENE DE TROIE 1916 Helicon
      - **LA TROIENNE 1926 Teddy** founded a family p. 137
      - **ADARGATIS 1931 FO Asterus**
        - **ARDAN 1941 FD, GP Pharis** won 13 other races including the Coronation Cup. Exp. USA 1949 Sired *Hard Sauce.
        - **and his brother**
        - *PARDAL 1947 imp. UK '51 **1947 Pharis** won Jockey Club Stakes and 5 other good races. p. 151

and her sister

Mated with: – MELTON

- 1903 ABSURDITY won 2 mod. races bred 8 wnrs. of 26 races – £38,000
  - **JEST 1910 1,O, Sundridge**
    - **HUMORIST 1918 D. Polymelus** died of tuberculosis soon after the Derby.
    - LAUGHTER 1919 Pommern did not win bred 5 wnrs.
      - ENSOLEILLEE 1924 Sunstar (exp. Fr '30) won 1 small race.
        - **SIGNAL LIGHT 1936 Pharos** Inbred Polymelus and Sundridge 3 & 4 rems. won 2 races, Sired the dams of Homeward Bound O and Democratic.
  - **BLACK JESTER 1911 L, Polymelus** sired dams of *Barneveldt GP and Sir Walter Raleigh
  - GESTURE 1918 Sunstar won 1 good race
    - AMUSE 1927 Phalaris did not win
      - **PICTURE PLAY 1941 1, Donatello II** bred 6 wnrs.
        - QUEEN OF LIGHT 1936 Borealis won 3 races, bred 6 wnrs.
          - CRYSTAL PALACE 1956 Solar Slipper won 5 races and £5,800
            - **ROYAL PALACE 1964 2, D, E, GVI, Ballymoss** sired DUNFERMLINE 1974 O, L.
          - **PICTURE LIGHT 1954** Court Martial
            - **WELSH PAGEANT 1966 Tudor Melody** ran 20 times in 4 seasons, won 11 inc. Free h'cp. Became top-class miler. Sire of Swiss Maid 1975 Ch.

his most famous daughters, Lady Josephine and Sun Worship, the *element of speed,* which he seems to have introduced just when it was needed, had very widespread repercussions. He was unsound in the wind. His son ***†SUNSTAR** 1908 2, D, when mated with Absurdity, started a very promising line with Gesture. Sunstar's dam was the unashamed selling-plater **Doris** who bred 10 winners including Price Dorrie 1911, 1, O.

FURZE BUSH 1891

The origins of **FURZE BUSH** 1891. By **FITZ JAMES** 1875 who appears not to have run. By the stout GC winner, *†The Scottish Chief from a very well connected bottom line, his name is to be found elsewhere only as sire of the 2nd dam of Slieve Galleon 2.

Ex BRIER BUSH 1884	by the not very notable **Tynedale**, a stayer. Brier Bush won 2 races, exp. USA 1892.
Ex BRIONY 1878	by Adventurer.
Ex MAID OF THE GLEN 1856	by Kingston.
Ex GLENGOWRIE 1851	by Touchstone. Founder of 3 classic lines, ancestress of We Shall Remember, see later p. 25. Family no. 1.

Inbred to Newminster at the 4th remove and with his sire Touchstone once in the 4th and 3 times in the 5th, the genetic causes of Furze Bush's eventual success are obscure, unless this background has anything to do with it.

Furze Bush ran 58 times and won 7 races at 5f.

She had 12 living foals, each one by a different and sometimes obscure stallion; it can be no coincidence that the 3 matings which brought about her only successful daughters were with fairly well-known horses.

FURSE BUSH'S MATINGS:

GALEAZZO 1893, a middle-distance horse by Galopin, beautifully bred. He had already visited France as a 3yo and returned; he sired later the 1000 gns. winner Atmah and was exp. Italy 1909.

PIETERMARITZBERG 1898, a stayer (Jockey Club Stakes), was also impeccably bred, being by St. Simon from an Isonomy mare. He left little else behind in this country, having been sent to Argentina in 1906.

Sundridge v.s. p. 11.

FURZE BUSH 1891

Mated with:– GALEAZZO

Produce:– 1901 WHINBLOOM won 7 races, bred 2 wnrs.

LUPERCALIA 1914 Lycaon did not win

CANTALUPE 1920 Amadis won 5 races

ORAMA 1932 Diophon won once, bred 11 wnrs.

BEAUSITE 1933 Bold Archer won 3 races bred 10 wnrs.

HOLWOOD 1951 Umidwar won 4 races

LAKEWOODS 1958 exp. USA '69 Hyperion did not win

WATERLOO 1969, 1, exp. USA '72 **Bold Lad**

RESPITE 1941 Flag of Truce won once bred 8 wnrs. of 21 races £37,000

NEARULA 1950 2, Ch. Nasrullah died aged 10 having achieved nothing of note at stud.

Mated with:– PIETERMARITZBERG

1905 VELDT won twice bred 3 wnrs.

MARVELDT 1914 Marcovil did not run, exp. France 1921

MACKWILLER 1923 Verwood (FI) & 5 other races.

MAMMEE 1929 Bruleur won twice

MIGNON 1935 Epinard

MIMI 1943 Black Devil

VIMY imp. UK '55 **1952 GVI, Wild Risk** exp. Japan 1965 sired Khalkis 1960 E,

MESA 1932 1, Kircubbin Her sire *†KIRCUBBIN 1950 (Ir.L), by Captivation from a mare by Hackler, sired also *† Chateau Bouscaut FD, (gd-sire of Pinza D,) and the dam of Eclair Au Chocolat AT.

HISTORIC 1939 Solario won 6 races a stayer and sire of stayers: sub-classic.

and his sister

SOLESA 1940 Solario won once

†RATIFICATION 1953 exp. Jap '65 **Court Martial** won 6 races inc. Coventry and Richmond stakes and the Greenham. Sired the dam of Sassafras FD, AT.

Mated with:– SUNDRIDGE

1907 CATALPA won twice, bred 2 wnrs.

BAY BERRY 1913 Bayardo won once, bred 4 wnrs.

***†BERESFORD 1921 Friar Marcus** sprinter, won 5 races at 2y and one (6f) at 4y. Sire of †Portlaw v.s. p. 14

SATIRICAL 1891

The origins of **SATIRICAL** 1891. By the middle-distance †**SATIETY** 1885, a well bred horse by Isonomy from a mare by Cremorne. He was sent to Germany in 1899.

Ex CHAFF 1880	by the Irish **WILD OATS** 1860 who appears not to have run and is of little other renown. Chaff did not win.
Ex CELERRIMA	by STOCKWELL.
Ex SLANDER	by PANTALOON
Ex BANTER	by Master Henry. Dam of *Touchstone L, GC2, and his brother Launcelot L, and gd-dam of *†Macaroni 2, D, and Satirist L: her daughters started no less than 4 branches of the prolific family no. 14.

Satirical, despite her not very distinguished parents, had a pretty strong pedigree, she was inbred to Stockwell at the 3rd and 4th removes.

She had 4 other foals by well-known sires, however, these two visits to the mighty St. Simon did not take long to yield classic results. *†**ST. SIMON** 1881 GC was sired by the Derby winner Galopin from St. Angela by King Tom; she was 16 when this prodigy was born but she had hitherto produced nothing of merit and possibly that was the reason why St. Simon was entered only for the 2000gns. and not for the Derby or Leger. Anyway his owner died during his 2yo career and, according to the current rules, his entry in the Guineas became void. Never beaten, he won 10 races at 2 and 3 years, including the Gold Cup by 20 lengths as a 3yo. It is said that his jockey on that occasion could not pull him up for another mile. Although sound, he did not run as a 4yo and did not go to stud until he was 5. Champion sire of winners 9 times and of brood-mares 6 times, he sired the winners of 25 of the races some of which I miscall classics in this book. He lived to be 27.

SATIRICAL 1891

and her two St. Simon Offspring

Mated with: – ST. SIMON

Produce: –

1900 *† RABELAIS
won from 5 to 21f, exp. France 1904 where he sired, among many good animals, the winners of 4 classics, *†Havresac II (sire of Nogara – see later) and *†Rialto. He heads two lines of St.Simon's dynasty, leading to Ribot and Worden. Rabelais lived for 29 years.

and his sister

ST. SIMON

1904 SIMONE
won once at 2y, bred 1 wnr.

FAIR SIMONE
1917 Farman
exp. France 1921
won 7 races

†FINGLAS
1923
Bruleur
a stayer of renown, sired the dams of *Vieux Manoir GP, Fine Art and Goyama.

SKY GLORY
1922
Sky Rocket
did not win
bred 6 wnrs.

EMPIRE GLORY
1930
Singapore
won one good race

NOORANI
1950
Nearco
won one good race

***SHESHOON**
1956 GC
Precipitation
won 7 other races also and was 2nd by a short head in the 20f Cadran. His average winning distance was 13.8f. To stud in Ireland 1961 and has sired 3 classic horses: – Sassafras 1967 FD,AT, Pleben 1969 GP and Mon Fils 2.
see also p. 175

***CHARLOTTESVILLE**
1957 FD, GP, (Lupin)
Prince Chevalier
died 1972. Sired Charlottown 1963 D

WE SHALL REMEMBER 1891

The origins of **WE SHALL REMEMBER** 1891. By ***†ISONOMY** 1875 GC2, in addition to his reputation as a great stayer, he would have been known by this time as the sire of Seabreeze O, L. However, he was destined for greater things and it will be seen later, how, apart from having sired Gallinule and Isinglass, he became a powerful link in the dynastic chain leading to Persian Gulf, Vieux Manoir and Crepello.

Ex SONSIE QUEEN 1875	by the long-range stayer **MUSKET** 1867 who, exp. N.Z. 1878, became the sire of the mighty Carbine who is worshipped there to this day. Had this book been started pre-St. Simon, she would have merited a page to herself; she did full justice to her stout ancestry by producing, from another Isonomy daughter, a line leading to Cherimoya O, *†Cameronian 2, D, Ch., and *The Macnab. Another daughter made her the ancestress of Galvani Ch. and Cerisoles 1954 FO.
Ex HIGHLAND LASSIE	by STOCKWELL.
Ex GLENGOWRIE	by TOUCHSTONE, see p.1. Family no. 1.

We Shall Remember did not run; she was inbred to Stockwell at the 3rd remove.

She was exported to France 1903, and had 4 foals there, making 9 in all.

WE SHALL REMEMBER'S MATINGS:

JOHN MORGAN 1887, a 6 to 8f handicapper, stud fee £5. By Muncaster from a Springfield mare, his daughter Trust was endowed with strains of Stockwell to the extent of 3 at the 4th remove and one at the 5th.

†LAVENO 1892 Ch. was a bit more expensive, he was by Bend Or from the dam of †Orvieto by Macaroni. He sired later Galvani Ch., *Valens and the winner of the 1917 Grand National, Ballymacad. Membrilla (p.27) was likewise inbred to Stockwell 3 times at the 4th rem. where she also had 2 strains of Thormanby, Laveno being inbred to that horse at 3rd rem.

WE SHALL REMEMBER 1891

whose influence seems now to be confined to USA and France

Mated with: –

Produce: –

JOHN MORGAN

1901 TRUST
did not win,
bred 8 wnrs.

SECURITY
1912
Simon Square
won 6 races
bred 4 wnrs.

DANDELION
1921
Rochester
did not win

WEEDS 1926
Arion
won 8 races

GOOD DEAL
1932
Apelle
won 7 races
bred one wnr.

SALECRAFT 1938
Orpen
did not win

SYLKO 1948
Nearco

MIGOLETTE
1955 Migoli

NAUJWAN 1960
Ommeyad

GIACOMETTI
1971 Ch.
Faberge II
Inbred Nearco
4 rem. Exp. USA
1975

***†STRAIGHT**
DEAL 1940
D Solario
Sired the dams
of *Doutelle,
†Above Suspicion
and *Silly Season.

ARIADNE
1922
Arion

YASNA 1936
Dastur
won 4 races
bred 9 wnrs.
£34,900

***DARIUS 1951**
2, E, Dante
sired *Derring Do & Pia
O. His sire *†DANTE
1942 war-time D and
his brother *SAYAJIRAO
1944 L, were by Nearco from
the Dark Legend mare, Rosy
Legend. Dante sired also
Carrozza O, Toulouse Lautrec
and the dams of *Tudor
Melody, Pieces of Eight Ch.
and Bolkonski 2. Sayajirao
sired Gladness GC, Indiana L
and the dam of Irish Ball IRD.

LAVENO

- 1902 MEMBRILLA
  exp. France 1908
  won 2 small races
  - HALLEBARDE 1901
    The Admirable Crichton
    - CANALETTE 1922
      Cannobie
      won 4 good races,
      bred 7 wnrs.
      - BADOLONA 1929
        Badajoz
        won 6 races
        - BALACLAVA
          1936
          Caligula
          - BARINA 1951
            Balthazar
            - **BLA BLA**
              1962 FO
              Lavendin
              Inbred Epinard
              3 & 4 rems. &
              Badajoz 4th.
      - NICA 1934
        Nino
        won 7 races,
        bred 8 wnrs.
        - **MONTENICA**
          1944 FO
          Djebel
          bred 9 wnrs.
        - and sis. DJEBELLICA
          1948 (Ir. O)
          Djebel
          - DJEBEL IDRA
            1957
            Phil Drake
            - ***BON MOT III**
              1963 AT
              Worden
              sired Lassalle GC
              see p. 142
            - TIDRA
              1964
              Prince Taj
              - **TENNYSON**
                1970 GP
                Val de Loir
      - and her brother **CANOT 1935**
        Nino
        won 4 good races
        inbred Isinglass
        4 rem. Sired
        Imprudence 1,O,
        (FL)
      - MA CANAILLE
        1937
        The Macnab
        did not win
        - **GOOD LUCK 1946**
          FD Goya II
          exp. Brazil
  - MEDIA LUNA II
    1913
    Northeast
    - DIEUZE 1918
      Go To Bed
      - **DAVOUT 1933**
        (F2) Parth
        Inbred Perth II 3 & 4
        rems and Hampton 4th.
        Sired *Eblè. His sire
        PARTH 1920 AT & 6
        other races, by Polymelus
        from a William The Third
        mare, was inbred to
        Hampton at the 3rd rem.
        Exp. Fr '25.

ISOLETTA 1891

The origins of **ISOLETTA** 1891. By ***†ISONOMY** 1875 GC twice, Goodwood & Doncaster Cups also the Cambridgeshire, with which in view he was kept in the dark and did not run in the 3yo classics. It is said the winnings were tremendous. He was a small horse by Sterling from a Stockwell mare.

Ex LADY MUNCASTER 1884	won 8 races, by the miler ***†Muncaster** 1877.
Ex BLUE LIGHT	by Rataplan.
Ex BOREALIS	by Newminster
Ex BLINKBONNY 1854 D, O	by Melbourne. The dam also of the brothers by Stockwell, *†Blair Athol D, L and †Breadalbane.
Ex QUEEN MARY 1843	by Gladiator who did not win but had 17 foals in all of which no less than 6 daughters founded classic lines. Family no 10.

Isoletta did not run; she was inbred to Stockwell at the 3rd & 4th removes and the background of her pedigree is remarkable for the fact that the scion **BIRDCATCHER** 1833 appears once in the 4th generation, 3 times in the 5th and once again in the 6th. One wonders idly how many of the animals in Isoletta's pedigree inherited Birdcatcher's renowned grey or white specks.

She was sent to Italy in 1903 after having had 7 foals.

ISOLETTA'S MATINGS:

***†GALOPIN** 1872 D, was 3 times champion sire of both winners and broodmares. Sire of St. Simon and the 2000 winner, Galliard, Galopin was inbred to Voltaire at the 3rd remove, which of course gave Galicia 2 of that strain at her 4th.

***†FLORIZEL II** 1891, one of the 3 great brothers by St. Simon from Perdita II, was a staying Cup horse and he sired 2 classic winners. His daughters bred a GC and a 2000 winner also the great French horse, *†Sardanapale FD, GP.

ISOLETTA 1891

Mated with:–

Produce:–

GALOPIN (St. Simon's sire)

1898 GALICIA
won once at 2y
bred 4 wnrs. of
42 races and
£88,000

***†BAYARDO 1906**
L, GC, E, Ch.
Bay Ronald
won 22 out of 25
starts; became
champion sire twice
in the short time
before he died aged
11, an incalculable
loss. However, he
got *†Gainsborough
2, D, L, GC who
carried on the
Hampton dynasty
for him. He was
inbred Galopin 2
and 4 rems and
Sterling 4th.

***†LEMBERG 1907**
D,Ch.(twice), =E.
Cyllene
won also Doncaster
Cup, Coronation Cup
and Jockey Club
stakes. Great Racehorse
though he was, he did
not carry on the
patriarchal line of
Cyllene, this, as will
be seen later, was left
to Polymelus. Lemberg
was inbred to Isonomy
at 3rd rem. He sired 4
classic winners and his
daughters bred the
winners of 6 classics
and the swift stallion
Denturius.

SILESIA 1909
Spearmint
did not win
bred 4 wnrs.

MY DEAR 1915
O, Ch.
Beppo
her sire, *†BEPPO
1903 won the
Hardwicke and
Jockey Club Stakes
and was 3rd in the
Leger by two heads;
He sired also Aleppo
GC and the dam of
†Gay Crusader 2,D,
L,GC,Ch. Beppo
was by Marco from
a St. Frusquin mare.

CICERONETTA
1916 Cicero
won twice
bred 5 wnrs.

CONCORDIA
1929
Son In Law
won 1 good race
bred 4 wnrs.

PICAROON 1922
Ch. Beppo
won 7 other races, died
aged 4.

NAMESAKE
1944
Fearless Fox
did not win
exp. France 1947

NAMAGUA
1954
Fontenay

NELCIUS
1963 FD
Tenareze
inbred Tourbillon
4 rem. imported
UK 1967,
exp. Japan 1971

FLAG OF TRUCE
1934 Truculent
a miler, inbred Bay
Ronald 4 rem.
sired d. of Nearula 2, Ch.

FLORIZEL II (St. Simon's son)

1903 CHERE REINE
did not win
bred 3 good winners.

ALEPPO 1909 GC
Beppo
a great stayer, inbred
St. Simon 3 and 4 rems.
exp. Belgium 1921.

ADMIRATION 1892

The origins of **ADMIRATION** 1892. By *†**SARABAND** 1883, a miler, inbred to Touchstone at the 4th rem. Would not yet have made his name as a sire. Exp. Germany 1896.

Ex GAZE 1886	by the blood-vessel breaker **Thuringian Prince** 1871, won Hunt Cup, not very distinguished, but sired a GC winner. Gaze did not run.
Ex EYE-PLEASER 1876	by Brown Bread.
Ex WALLFLOWER	by Rataplan.
Ex CHAPERON	by Flatcatcher, family no 14.

By a not very renowned sire, from a useless mare by a b-v breaker, Admiration hardly looked like making history. Nevertheless her advent can only be described as an explosion. She came from a line of ineffective though well-bred females, 6 of them in 55 years and not a trace of a decent race or winner between them. It was from **BOADICEA** 1807 that this family emerged into classic realms; her first foal came in 1814, then she had some seasons hunting and in 1820, 21 and 26 produced 3 family-founding daughters.

Admiration ran 18 times in 4 seasons on the flat, winning 2 tiny races, nor was she a success under Irish N.H. Rules.

She had 13 foals in just that number of years.

ADMIRATION'S MATINGS:

Laveno (p. 25), Gallinule (p. 13), Isinglass (p. 13).

The Laveno union gave Veneration 3 strains of Thormanby in her 4th generation.

ADMIRATION 1892

Mated with:– LAVENO; GALLINULE; GALLINULE

Produce:–

- LAVENO
  - 1900 VENERATION II
    a winner herself, bred 4 wnrs. of 19 races and £18,500.
    - SACRED IBIS 1908 Gallinule did not win
      - INCENSE 1917 Roi Herode won once exp. France 1924
        - SIMONE VERGNIS 1923 Diadumenos won once
          - CYBIANE 1930 Blandford exp. France 1931 won 2½ races. ret. 1935
            - **WHY HURRY 1940 O Precipitation** bred 7 wnrs.
          - MYRTLE GREEN 1937 Trigo visited France 1947
            - MARIEBELLE 1948 Mieuxce foaled France
              - **MARGUERITTE VERNAUT 1957 Ch. Toulouse Lautrec.**
    - **CRAGANOUR 1910 (disq. D) Desmond** a very good 2yo exp. Argentina 1913
- GALLINULE
  - **1901 PRETTY POLLY 1, O, L, Ch.** family of her own, p. 73
- and sis.
- GALLINULE
  - 1902 ADULA won £3,080 died young
    - MOUNTAIN MINT 1909 Spearmint
      - BRIDGEMOUNT 1919 Bridge of Earn won 4 races bred 6 wnrs.
        - **CRESTA RUN 1924 O Hurry On**

ISINGLASS

1903
THE ADMIRABLE CRICHTON
exp. France 1908
not very notable,
sired 2nd d. of
*Canot and 3rd d.
of Montenica FO
and Good Luck
FD.

GALLINULE

1905 MIRANDA
another sis. to
Pretty Polly
won twice
bred 7 wnrs. of
15 races and
£10,500

MIRAWALA 1923
exp. Fr '34
Phalaris
won 4 races,
bred 6 winners

BECTI 1929
Salmon Trout
exp. France 1932

SIND 1933 by Solario and his sis.
won 2 good races, exp. France
1936, Argentina 1940, sired d. of
*†Worden, (Wash. Int)

STAFARALLA 1935
Solario won 2 big races,
bred 10 wnrs. of over
£22,000

***TEHRAN 1941 L**
Bois Roussel

BIBI SAHIBA
1934 Blenheim
won once

and her sis.

BOUILLABAISSE
1935 Blenheim
exp. Fr. '38
won twice at 2y

ARGUR 1949
E Djebel
exp. Argentina

CHINA MAIDEN
1938 Tai Yang
bred 5 wnrs.

EASTERN
GRANDEUR
1945
Gold Bridge
exp. USA 1956

ROSE OF INDIA
1955 Tulyar

VIRTUOUS
1962
Above Suspicion

HARD TO BEAT
1969 FD
Hardicanute
exp. France 1970

STAR OF INDIA
1953 Court Martial

BLUE SASH
1958 Djebe

HANINA 1965
Darling Boy

MRS. McARDY
1974 1
Tribal Chief

BLACK CHERRY 1892

The origins of **BLACK CHERRY** 1892. By †**BENDIGO** 1880 E, Ch., a tough Irish horse which stood up to 6 seasons racing but was not a success as a stallion other than getting Black Cherry. Stood in Germany 1895-1900.

Ex BLACK DUCHESS 1886	by *†GALLIARD 1880 2. Black Duchess won once, dam of *†Bay Ronald see p.149, 172, App. C, App. D.
Ex BLACK CORRIE	by STERLING, exp. France 1886
Ex a bay mare 1861	by WILD DAYRELL.
Ex LADY LUREWELL	by Hornsea and 3 gens. back to AMAZON 1799, ancestress of Pocahontas. Family no. 3.

Black Cherry, a first foal, ran only once, winning a small seller as a 2yo.

She had 13 foals to live, and produced 9 winners of 41 races and £30,500 including a 1000 gns. and Oaks winner. She is also 5th dam of the 1947 Grand National winner, Caughoo. Her matings now introduce two important sires: *†**AYRSHIRE** 1885 2, D, E, by Hampton from the great Galopin mare, **Atalanta**. He made his name chiefly as a sire of brood-mares of which 5 bred classic winners, he sired also the Oaks winners, Airs And Graces and Our Lassie. Of his sons, *†Symington proved the best of four quite useful sires. Jean's Folly was inbred to Galopin at the 3rd and 4th removes. †**WHITE EAGLE** 1905, a middle-distance horse (though 2nd in a moderate Leger) by Gallinule (by Isonomy) from a Galopin mare and a granddam by Isonomy. This packs Blanche's pedigree with 2 Isonomys and 2 Galopins in the 3rd and 4th gens. respectively, also 2 Sterlings in the 4th. White Eagle got many very great mares besides Blanche: Dolabella and Lady Peregrine, to mention only two. Blanche was one of the original mares at The National Stud, presented to the nation by Lord Wavertree in 1914. She was 17 when she conceived Nun's Veil to the King's sprinter, Friar Marcus.

BLACK CHERRY 1892

Mated with: – AYRSHIRE | ISINGLASS | WHITE EAGLE

Produce: –

AYRSHIRE: 1901 JEAN'S FOLLY won 6 races, bred 10 wnrs. of 24 races £14,000

- **NIGHT HAWK 1910 L** **Gallinule** exp. Australia 1918
- WHITE LIE & her sis 1911 White Eagle won once
  - WHITE FOLLY 1924 Swynford did not win bred 4 wnrs.
  - JOY RIDE 1931 Buchan won once
  - GRANDE CORNICHE 1941 Panorama won 3 races bred 6 wnrs.
  - BEBE GRANDE 1950 Niccolo dell'Arca won 8 sprints and £18,500, bred 4 wnrs.
  - BABY DOLL 1956 Dante won twice at 2y
  - **PIECES OF EIGHT 1963 E, Ch.** exp. USA '67 **Relic** see p.178, App. C, App. D.
- ARCTIC NIGHT 1920 White Eagle
  - LADY OF THE SNOWS 1928 Manna
  - LIFE HILL 1940 Solario
  - CHAMBORD 1955 Chamossaire
  - ***CARO 1967 (F2)** exp. France 1968 **Fortino II** Prix Ganay (10½f) his sire's av. wins were 6.3f and he was by Grey Sovereign, so, since he has sired CRYSTAL PALACE 1974 FD and MADELIA FO, one assumes that the staying potential comes from the female line thanks to Chamossaire, Solario and Manna, not to mention, Crystal Palace's dam Hermieres FO by Sicambre.

ISINGLASS: **1902 CHERRY LASS 1, O.** died young, having bred two winners.

WHITE EAGLE: 1912 BLANCHE did not win, bred 6 wnrs. of 20 races and £16,000

- ***†BLANDFORD 1919 Swynford** won 3 out of 4 good middle-distance races. Inbred to Isonomy at the 4th rem. Champion sire 3 times, he got the winners of 12 classics including Blenheim D, his successor in the Isonomy stirp. His daughters accounted for the winners of 6 classics. His sire, *†SWYNFORD 1907 L,E, whose tremendous achievements are set out on p.38, was lucky to survive a broken fetlock when he was 4; he was sired by the bad mannered and still worse racehorse, John O'Gaunt out of the Oaks winner Canterbury Pilgrim who is my next study.
- NUN'S VEIL 1930 Friar Marcus won 3 races
  - CLARENCE 1934 Diligence did not run, bred 4 wnrs.
    - SISTER CLARA 1938 Scarlet Tiger did not run bred 7 wnrs.
      - AUNT CLARA 1953 Arctic Prince did not win, bred 5 UK wnrs.
      - **SANTA CLAUS 1961 D, IRD.** **Chamossaire**
    - **SUN CHARIOT** & her sis. **1939 1, O, L.** **Hyperion**
      - **BLUE TRAIN 1944** **Blue Peter** inbred Chaucer 4 rem. sired d. of *Kashmir II 1963 2.
    - CALASH 1944 Hyperion won once bred 3 wnrs.
      - **CARROZZA 1954 O** exp. USA '65 **Dante**
      - CARROMATA 1965 St. Paddy
      - **MATAHAWK 1972 GP** **Sea Hawk II** inbred Hyperion 4 rem.

CANTERBURY PILGRIM 1893 O

The origins of **CANTERBURY PILGRIM** 1893 O. By ***†TRISTAN** 1878 GC, Ch. (three times) a bad tempered and versatile horse who won the 6f July Cup one year and the GC the next. Stoutly bred by Hermit from a Stockwell mare. Exp. France 1884 -91, sent to Austria 1894.

Ex PILGRIMAGE 1875 2, 1	By ***†THE PALMER** 1864 who ran 26 times and won 7 mid-distance races. Exp. Germany 1876. Sired also Jenny Howlet O and her sister Palmflower see p.131.
Ex LADY AUDLEY	by MACARONI.
Ex SECRET	by MELBOURNE, a great granddaughter of Problem 1823 1, ancestress also of Phalaris. Family no. 1.

Canterbury Pilgrim had 10 living foals in all, among them 7 winners of 30 races value £37,500. Despite the mighty Swynford her main glory is derived from *all* her stallion sons, especially Chaucer who (note without the aid of Phalaris) also sired Hyperion's dam, Selene.

CANTERBURY PILGRIM'S MATINGS: ST. SIMON (p. 23)

***†JOHN O'GAUNT** 1901, won only one moderate race, nevertheless he was 2nd in both the 2000 gns. and the Derby, in the latter ridden by an amateur. He broke down before the Leger. Three things were remarkable about him – his bad manners which marred his racing career and stud duties, the fact that both his parents won the GC and his production of Swynford. He also got Kennymore 2. John O'Gaunt was by Isinglass from the fabulous La Fleche by St. Simon (p. 167).

CANTERBURY PILGRIM 1893 O.

Mated with:– ST. SIMON

Produce:–

1900 CHAUCER ran 35 times, won the Gimcrack and 7 other races, av. winning dist. 8.3f Famous as a sire of mares, particularly when they were mated with Phalaris; this combination produced the winners of 9 classics.

and his sis.

Mated with:– ST. SIMON

Produce:–

WIFE OF BATH 1904 did not win bred 4 wnrs.

WHITEWASH 1910 White Eagle won twice, bred 5 wnrs.

BLANCHISEUSE 1923 Phalaris won 2½ races

WHITEBAIT 1931 Sansovino did not win

NONATS 1937 King Salmon did not win

DOUBLE DEAL II 1946 Straight Deal won 3 races Exp. USA 1952

SILLY SEASON 1962 Ch. Tom Fool foaled USA, imp. UK 1963. His sire *†Tom Fool 1948 a US miler, won 10 out of 10 starts; sire also of the dam of Meadow Court IRD, GVI.

ORANGE FLASH 1951 Court Martial did not win bred 4 wnrs.

NOW WHAT 1956 Premonition won 3 races at 2y. bred 4 wnrs. in UK

MON FILS 1970 2, exp. Fr. '75 **Sheshoon** closely inbred to Precipitation at 2nd and 3rd rems. also to Fair Trial at 4th.

Mated with:– JOHN O'GAUNT

Produce:–

***†SWYNFORD 1907, L, E.** won also 6 other races. As ancestor of Blenheim, Donatello II and Crepello, he played such an immense part in the creation of classic winners in Europe that I'm going to list his main stud achievements: Four winners of the 1000, including Saucy Sue 1, O (dam of Truculent), two Leger winners including Keysoe (3rd d. of Ballymoss), *†Sansovino D, †Hainault, †Schiavoni, *Stratford, Under Fire, Rose Red (p.133) and his successor in the Isonomy line, *†Blandford. His daughters bred Miracle E, Tideway 1 (d. of Gulf Stream), *†Tornado, Sun Stream 1, O, †Victrix, Truculent, †Warrior, Heliopolis and Seven Seas.

and his less important bro.

Mated with:– JOHN O'GAUNT

Produce:–

1910 HARRY OF HEREFORD never won, exp. France 1920. Sired the 2nd dam of the American champion *†War Admiral.

Mated with:– ST. FRUSQUIN

Produce:–

1905 PILGRIM'S WAY

WAYFARER 1919 Call of the Wild

RIGHT OF WAY 1925 Craig an Eran

PRIVATE ENTREE 1933 Prestissimo

SKI CLUB 1950 Hook Money 1951 sprint h'cpr. ran 37 times, won 9.

HORNTON GRANGE 1967 Hornbeam

SWISS MAID 1975 Ch. (3y) Welsh Pageant

CHELANDRY 1894

The origins of **CHELANDRY** 1894 1 and 2nd in both Oaks and Leger. By the superbly bred but, as far as racing was concerned, totally useless †GOLDFINCH 1889. He was unsound in the wind and exp. USA 1894. By Ormonde from **Thistle** who later became dam of the 2 winners of 4 classic races, ***COMMON** 2, D, L and Throstle L.

Ex ILLUMINATA 1877	by *†ROSICRUCIAN 1865, a long distance stayer, by the time of this mating he would have been known as the sire of 2 classic winners; besides Chelandry his daughters bred the winners of 4 classics. Illuminata won 1 good ½mile race at 2y. dam also of *†Ladas 1891 2, D, and daughters who founded 3 classic families.
Ex PARAFFIN	by Blair Athol whose daughters founded 3 classic branches.
Ex PARADIGM	by Paragone. D. of *†Lord Lyon 2, D, L & of his sister Achievement 1, L.
Ex ELLEN HORNE	by Redshank.

Four of the greatest mares in racing history led up to a 5th in Chelandry, in addition to Chelandry's achievements overleaf, these 4 were ancestresses of 26 classic winners, not to mention the important sires, Roi Herode and Prince Palatine. Family No. 1.

Chelandry had 16 foals and bred 6 winners of £42,600.

HER MATINGS: (not previously discussed)

***†ST. FRUSQUIN** 1893 2, E, by St. Simon from a less aristocratic mare, was unlucky to have been born the same year as the wonder-horse Persimmon, whom he actually beat as a 2yo, but to whom he was 2nd by a neck in the Derby. He sired the winners of 7 classics and the dams of the winners of 8.

***MARCO** 1892, a moderate handicapper, won the Cambridgeshire at 3 y car. 7.9. Sired also *†Beppo, †Omar Khayyam, *†Marcovil and Sprig, who won the Grand National in 1927.

WILLONYX 1907 GC also Cesarewitch car. 9.5, a grand stayer, but the non-winning Bobolink was his only achievement at stud.

CHELANDRY 1894 1.

Mated with: – AYRSHIRE

Produce: – 1900 SKYSCRAPER first foal won £4631, bred 5 wnrs.

ASCENSEUR 1911 Eager did not win, bred 4 wnrs.

SUNSTEP 1916 Sunstar exp. USA as a foal

GALADY II 1927 Sir Galahad won 8 races (7 in USA), bred 5 wnrs.

GALATEA II 1936 1, O, Dark Legend

BOREALE fld.Fr. exp. USA 1938 Vatout won once, bred 6 wnrs.

SINGING GRASS 1940 War Admiral won 7 small races

***NEVER SAY DIE 1951 D, L, Nasrullah** foaled USA, imp. as a yearling; to stud at 4y. His generous American owner later presented him to the National Stud with the wish that 25% of the nominations went to Irish breeders. Sire of Never Too Late 1,O, Larkspur D and *Immortality.

ISINGLASS

1902 SAMPHIRE did not win, bred 6 winners.

WRACK 1909 Robert Le Diable a miler, exp. USA 1915. Sired Petee Wrack and the dam of Flares GC, Ch.

JUNIATA 1916 Junior won once

JIFFY 1931 Hurry On did not win bred 8 wnrs.

OCEAN SWELL 1941 D, GC, Blue Peter possibly Blue Peter's best son. Sired †Fastnet Rock.

ST. FRUSQUIN

1905 POPINJAY won one good race at 2y, bred 8 wnrs. £15,900

GOOD AND GAY 1912 Bayardo won 2 big races, bred 4 good wnrs.

SAUCY SUE 1922 1, O, Swynford inbred St. Simon at 4th rem.

TRUCULENT 1928 Teddy a miler, inbred Bay Ronald 3 and 4 rems. Sired Flag of Truce

POPINGAOL 1913 Dark Ronald bred 11 wnrs. of 27½ races £55,900

POGROM 1919 O, Lemberg inbred Isonomy 4 rem.

BOOKLAW 1924 L, (exp. Fr. '32) **Buchan**

RHODES SCHOLAR 1933 E, Pharos & his sis. won also St.James Pal. Stakes. Exp. USA 1940 having been fld Fr. and imp. '34. Sired Black Tarquin L.

LIGHT SENTENCE 1934 Pharos

NICCOLINA 1950 Niccolo dell'Arca exp. France '58

CHAPARRAL 1966 GP, Val de Loir

MARCO

1907 †NEIL GOW
2, =E,
Died aged only 12, sired the dams of Rose of England O, *†Chateau Bouscaut FD, *†Rodosto 2, Chatelaine O, =Ch. *Bubbles & Cadum

WILLONYX

1913 BOBALINK
did not win
exp. USA 1916

ST. JAMES 1921
Ambassador IV
inbred St. Simon 4 rem. Grandsire of †Johnstown

SUNSTAR (p.20)

1914 PENNULA
did not win
bred 3 wnrs.

KILMENY
1921
Son In Law
won twice

QUINCE 1927
Pharos
did not win
exp. France 1939

LA TRUITE II
1936 Salmon Trout

LA TAGLIONE
1948 Colombo

***TENAREZE**
1953 Goyama
Inbred Colombo 2 and 3 rems. and to Salmon Trout 3 and 4. Sired Nelcius FD. His sire GOYAMA 1943 won 11 mid. dist. races, by Goya from a Finglas mare, inbred Bruleur 3 and 4 rems. Sired also the dam of Connaught E.

GAY BIRD 1923
Gay Crusader
did not win
exp. USA 1934

COLUMBA 1930
Colorado
exp. France 1935

***CLARION 1944**
Djebel
inbred Gay Crus. 3 rem. won 10 races in 3 seasons av. 6.6f
sired *†Klairon (F2) p. 179

and her sis.

BOOK DEBT
1925 Buchan
did not win,
bred 8 wnrs.

PAY UP 1933
2, Fairway
stood in France 1947 onwards, a sub-classic sire.

and her sis.

TOLBOOTH
1929 Buchan
did not win
exp. Italy
1932

TALMA 1933
Papyrus

***TRAGHETTO**
1941 (It.D)
Cavaliere
d'Arpino
sired Nuccio AT, (It.D)

FAIR COP
1933 Fairway
won twice
bred 7 wnrs.

FAIRLY HOT
1939 Solario
won 1 small race, bred 9 wnrs.

INDIAN NIGHT
1946 Umidwar
won 5 races,
bred 8 wnrs.

TANTALISER
1955 Tantieme
won 1 good race
bred 6 wnrs.

PROVOKE
1962 L
Aureole
exp. Russia 1966 and died very soon.

MATELDA
1947 exp. USA
'59 Dante
did not win

***TUDOR**
MELODY 1956
exp. USA '58,
ret. '60
Tudor Minstrel
ran only at 2y inbred Gainsborough and Pharos 4 rem. Sired *Kashmir II 2.

SIRENIA 1895

The origins of **SIRENIA** 1895. By ***†GALLINULE** see p.13 for an account of him – but his dam, Moorhen (by Hermit) must be noted here: **MOORHEN** 1873-91, at 3y ran 24 times winning 6 sellers; mated at 4 and ran 14 times without success. First foal at 5y, slipped at 6y; back into training when, aged 6, 7, 8 & 9 she ran in everything from flat sellers to hurdles and 'chases winning 20 times. Her daughter of 1887 was granddam of St. Victrix and her son, Pioneer, is also classically connected.

Ex CONCUSSION 1885	by †REVERBERATION 1871 who won up to 1m. Concussion did not run, dam of Llangibby 1902 E & Gddam of Spion Kop D.
Ex ASTWITH	by Wenlock.
Ex SISTER TO RYSHWORTH	by Skirmisher. *She is also Moorhen's dam.*
Ex VERTUMNA	by Stockwell. Family no. 19.

This breeding is incestuous in so far as both sire and dam had the same granddam. Stockwell appears twice in both the 4th and 5th generations.

SIRENIA won 9 races.

She had 9 foals altogether, 7 of which won 21 races and £28,500.

SIRENIA'S MATINGS: (not previously dealt with)

***†EAGER** 1894, a sprinting handicapper by the 2000 winner Enthusiast (by Sterling) which means that Scylla and her classic sister Electra had Sterling in the top lines of both their parents at the 3rd and 4th removes.

On a previous page I have mentioned Musket's son, **CARBINE** the N.Z. hero, where he won 33 races and spent 4 years at stud before being bought by the Duke of Portland in 1895 and coming to England where, in 1903, he sired ***†SPEARMINT** who, despite poor forelegs, won the Derby and went on to win the Grand Prix. We shall see more of him and his tough father chiefly through their wonderful daughters: Carbine's Sunshot and Miss Gunning, Spearmint's Plucky Liege, Seaweed and Catnip.

Spearmint's dam was **Maid of the Mint** by the huge Lord Lyon horse, ***†Minting GP**, and he was inbred to Stockwell at the 4th remove, which added to that scion's presence in Sourabaya's background.

SIRENIA 1895

Mated with: –

Produce: –

- EAGER
  - 1903 SCYLLA first foal did not win, bred 2 wnrs. of 13 races – £8,400
    - **JUNIOR 1909 Symington** inbred Galopin 4 rem. won 10 mid. dist. races, sired dam of Plack 1.
- ST. SIMON
  - 1905 SIBERIA won 5 good races bred 2 wnrs.
    - **SNOW MARTIN 1912 O,** (exp. Fr '23) **Martagon** *†MARTAGON 1887 won Gold Vase & Goodwood Cup. By Bend Or from a Macaroni mare. Sired also Musa O, Woolwinder L, †Buckwheat, Marlow and the dams of Mirska O, Kennymore 2, Montmartin FD, GP, and Prince Galahad
- EAGER
  - **1906 ELECTRA 1. sis to Scylla** bred 4 winners, inbred Sterling 3 and 4 rems.
    - SALAMANDRA 1913 St. Frusquin won twice, bred 7 wnrs.
      - FLYING SALLY 1919 Flying Orb did not win, bred 4 wnrs.
        - ***†EPIGRAM 1933 Son In Law** Great staying Cup horse; sired Souepi GC and the dam of Frieze O. Epigram was exp. Denmark 1954.
      - **SALMON TROUT 1921, L, The Tetrarch** eventually ('36) exp. S. Africa, sired King Salmon E.
      - DEVON LASS 1930 Hurry On did not win
        - MERRY DEVON 1937 Colombo bred 7 wnrs.
          - LIGHT-HEARTED 1944 Hyperion won once
            - COWES 1949 Blue Peter did not win, bred 4 wnrs.
              - ***HARD TACK 1955 Hard Sauce** won only at 2y. sired Right Tack 2, (Ir. 2)
    - **ORPHEUS 1917 Ch. (twice) Orby** won 11 races, sired Ormenos.
- CYLLENE
  - **1908 CELLINI** inbred Isonomy 3 rem. and both Hermit and Sterling at 4th. A good 2yo with little other distinction than having sired the 2nd d. of Scottish Union L.
- SPEARMINT
  - 1909 SOURABAYA (exp. Fr. '23) won once, bred 3 wnrs.
    - **COMRADE** (exp.Fr.'21) **1917 GP, AT Bachelor's Double** inbred Isonomy 4 rem. Sired 2nd d. of Cambremer L.

CONCERTINA 1896

The origins of **CONCERTINA** 1896. By ST. SIMON see p. 23.

Ex COMIC SONG 1884 — by ***†PETRARCH** 1873 2, L, GC by the Leger winner Lord Clifden from a mare by Orlando, he was inbred to Touchstone at the 3rd remove. His daughters bred the winners of 6 classics. He was sent to France aged 20. Comic Song did not run. exp. Russia 1898.

Ex FRIVOLITY — by MACARONI. Frivolity was a good 2yo.

Ex MISS AGNES — see p. 7. No. 16 Family.

Concertina did not run.

She bred 9 winners in all.

CONCERTINA'S MATINGS: (not previously described)

ROSELAND 1912, a moderate miler but, significantly, by the St. Simon horse William The Third GC, – line-breeding to the extent that Garron Lass had St. Simon at her 2nd and 3rd removes. It is hard to believe that such a reinforcement of this line did not play a part in the production of the fabulous Friar's Daughter, who, thanks to her sire's dam being by Persimmon, picked up yet another boost of St. Simon.

CONCERTINA 1896

and her immense influence in USA and France

Mated with:– SPEARMINT

Produce:–

1912 PLUCKY LIEGE
won £1811, exp. France 1930 after frequent visits. bred 10 wnrs. in both countries. Was named at first "Lucky Liege" which was changed in honour of the unavailing but heroic defence of that city against the Germans in 1914. She had 3 strains of Stockwell & 4 of Touchstone in the background of her pedigree.

***†SIR GALAHAD III 1920 Teddy**
a miler, stood 2 seasons in France, then exp. USA '26. Within the limited scope of European racing, he achieved at stud:– Gallant Fox (US triple crown & sire of Flares GC, Ch.), *† Black Devil (Don.Cup, mat.gdsire of Chamant FD, †Vimy GVI & Beau Prince II). He sired too the dams of Boswell L,E, Galatea II, 1, O, Black Tarquin L, Aurelius L, †Johnstown and †Mr Trouble.

and his bro.

***BULL DOG 1927 Teddy**
exp. from France to USA 1930 where he sired *BULL LEA (5 times US champion sire, got, among others, Hill Gail, †Bull Page, Mark Ye Well & Citation), Tiger and the dams of *†Tom Fool and Dark Star. Bull Dog died aged 27.

and their sister

NOOR JAHAN
1921 Teddy

- AVELLA
1936 Epinard
imp. UK 1940
bred 4 wnrs.
  - SATANELLA
1941
Mahmoud
did not win
exp. France
1951
    - PROMPT PAYMENT
1950 Rockefella
won 6f seller at 4y. Exp. Germany 1960
      - PESETA II
1957 Neckar
won twice
        - **PIA 1964 O. Darius**
    - SANELTA
1954 Tournament
      - **SANCTUS II 1960 GD, GP Fine Top**
inbred Blenheim 4 rem. His sire, *FINE TOP, won 16 races (av. 8.4f), was by Fine Art from a Vatellor mare, he sired also Fine Pearl FO, and Topyo AT.

***†ADMIRAL DRAKE 1931 GP, Craig An Eran**
won also 4 other good races in France sired *†Phil Drake D, GP, †L'Amiral and many big winners. His daughters bred Vattel GP, Hard Ridden D and *Turn To.

***†BOIS ROUSSEL 1935 D, Vatout**
inbred St. Simon 3 and 4 rems. see p. 173. imp. UK '38

ROSELAND *(St. Simon's grandson)*

1917 GARRON LASS
did not run
died aged 5.

FRIAR'S DAUGHTER
(exp. Fr. '35) 1921 Friar Marcus.
The only living foal of her dam, inbred St. Simon at 3rd and twice at 4th rems. Won 1 small race, bred 11 wnrs. of over £58,000

These mares of the Aga Khan's travelled regularly between UK and France

FILLE D'AMOUR
1926 Hurry On
won once

FILLE DE POETE
1935 Firdaussi
(Appx. A)
won 1 tiny race UK bred 1 other good wnr.

†THE PHOENIX 1940 (Ir. 2, D) Chateau Bouscaut
Sired the dam of Lorrenzaccio Ch. His sire, *†CHATEAU BOUSCAUT FD was a great stayer; by Kircubbin from a Neil Gow mare, he sired also *†Chanteur II (sire of Pinza D) and *Macaron.

FILLE DE SALUT
1928 Sansovino
won 4 races,
bred 4 wnrs.

FILLE DE SOLEIL
1935 Solario
did not run,
bred 8 wnrs.

***†SUNNY BOY 1944 Jock II**
inbred to Gainsborough 3 rem. Champion sire France, got Sun Cap O, Sica Boy AT, Tamanar FD and Midnight Sun (disq. AT and placed 2nd to St. Crespin) and the dams of *Val De Loir FD and Valoris O.

***†DASTUR 1929 =Ch. (Ir. D) Solario**
also won Coronation Cup, sired Umiddad GC, Dhoti and the dam of *Darius 2,E.

***†BAHRAM 1932 Blandford 2, D, L.**
Unbeaten in all 9 of his races, he earned over £43,000. His pedigree abounds with female greatness – in his 3rd gen. there is Black Cherry and Concertina and in his fourth La Fleche & Pilgrimage. He had a powerful St. Simon background. Before being sent to USA in 1940, he sired Turkhan L, *†Big Game 2, Ch. and *† Persian Gulf, also the dams of the winners of 2 classics. Apparently not a success in the States, he was sold to Argentina in 1946
see p.47, 176, App. C, App. D.

MISS GUNNING II 1897

The origins of **MISS GUNNING II** 1897. By **CARBINE** see p.43.

Ex MEMOIR 1887 O, L	by ST. SIMON who had 8 foals together, all by important stallions; Miss Gunning, however, was the only one to do any good.
Ex QUIVER 1872	by *†**TOXOPHILITE** 1855, a blood-vessel breaker who suffered also from a wind infirmity and started his stud career at 10 guineas. He is also the grand-sire of Carbine, so Miss Gunning is inbred to him at the 3rd remove. Quiver is dam also of Memoir's sister La Fleche 1889 1, O, L, GC, dam of John O'Gaunt.
Ex a bay mare 1860	by Young Melbourne
Ex BROWN BESS 1844	by Camel, an important foundation mare who appears twice in the 4th generation of Miss Gunning's pedigree. Family no. 3.

Miss Gunning II did not run.

MISS GUNNING II 1897

Mated with: –

Produce: –

CYLLENE

1905 SILENT LADY
won 3 good races
bred 7 small wnrs.

HUSH 1911
St. Serf
won 1 small race
exp. France 1915

UGANDA 1921
FO Bridaine
inbred St. Simon
3 & 4 rem.
her sire †BRIDAINE
was a long range
stayer whose dam
was by a son of
St. Simon.

UKRANIA 1926
FO Ksar
inbred Omnium II
3 and 4 rem.

UDAIPUR 1929 O
Blandford
won also 3 other
races bred 7 wnrs.

UMIDDAD 1940
GC Dastur
2nd by a hd in
Derby. Exp. France
1947.

UNA 1930
Tetratema
won once

***†PALESTINE**
1947 2, Fair Trial
sired Pall Mall 2.

***†UMIDWAR bro, to Udaipur**
1931 Ch. Blandford
won also Jockey Club Stakes
Sired Ujiji GC, *†Norseman,
Arbel and the dams of
Dynamiter Ch. (twice),
Festoon 1, *Counsel and
*Hardicanute.

BAY LADY 1915
Bayardo
did not win
bred 5 wnrs.

HURRY OFF 1925
Hurry On
did not win
bred 2 good wnrs.
exp. France '25.

HURRYLOR 1940
Vatellor
did not win,
bred 4 wnrs.

†L'AMIRAL 1947
Admiral Drake
won 5 good mid.
dist. races. Sired dam
of Gazala FO, (F1).

SWYNFORD

1914 ELISALEXI
did not win
bred 3 wnrs.
inbred St. Simon
3rd and 4th rem.

LOVELY CLOVELLY
1921 Clarissimus
did not win
exp. France 1925

LOVELY LEGEND
1936
Dark Legend

LEGENDE JOLIE
1942
Admiral Drake

VATTEL 1953 GP,
Vatellor

MAUND 1898

The origins of **MAUND** 1898. By **TARPORLEY** 1892, a St. Simon horse who ran and won as a 2yo only. He stood in France for 3 seasons.

Ex IANTHE 1887	by **The Miser** 1877, a well-bred Hermit horse which seems never to have run and was undistinguished at stud.
Ex DEVONSHIRE LASS	by Hampton.
Ex HIPPODROME	by OXFORD.
Ex HIPPODAMANIA	by Pelion.
Ex the foundation mare SLANDER	by Pantaloon. Her daughters founded 3 classic branches of Family no. 14.

Maund won 4 moderate races.

She had 6 other foals which did not achieve much.

MAUND'S MATINGS:

***†SYMINGTON** 1893, a good but not great racehorse, by Ayrshire from a St. Simon mare. As Ayrshire himself was out of a Galopin mare, Symington was inbred to Galopin at the 3rd rem. and his 2 daughters overleaf had St. Simon twice in the 3rd gen. and his father, Galopin, 3 times in the 4th.

MAUND 1898

Mated with: – SYMINGTON

Produce: –

1907 SCOTCH GIFT
won £1033
bred 7 wnrs. of
31 races – £31,500

***†TETRATEMA**
1917 2, The Tetrarch
inbred St. Simon 4
rem. Sired *Mr. Jinks 2,
Royal Minstrel E and
Four Course 1. The
only hope of survival
for The Tetrarch's
line is through
Tetratema's son
Bacteriophage to
Telerifique to ALIZIER
(Ap.A), but that
now seems forlorn.

sister

SYMINGTON

1912 LA MAULA
did not win
bred but 1 wnr.

LA MAURI 1922
Roi Herode
won 1 race
exp. France 1931
bred 3 wnrs. UK

LA BONI
1928
Pharos
exp. France
1931, ret.
UK 1937

BLUE BEAR
1932 (F1)
Blenheim

LE PAILLON
1942 AT,
Fastnet
later ran over
hurdles. Inbred
Pharos 2 & 3
rem.

and her sis.

LA LI 1935
Blenheim

TABRIZ 1947
Tehran
ran only at 2y.
won twice.
exp. France
1950. Sired
Taboun 2.

AIDETTA 1930
Phalaris
won 3 races

SABLA II 1937
Solario exp. Italy
1938

EVVIVA 1943
Navarro

OLA II 1949
Nakamuro

YRRAH Junr. 1964
Ribot
inbred Pharos 4 rem.
Imp. UK 1967,
exp. Canada 1969.
Sired HURRY
HARRIET 1970 Ch.

GLASALT 1898

The origins of **GLASALT** 1898. By the stayer ISINGLASS p. 13.

Ex BROAD CORRIE 1889	by the stayer HAMPTON p. 7.
Ex CORRIE ROY	by GALOPIN.
Ex CORRIE	by STOCKWELL.
Ex MAYONNAISE 1856 1	by Teddington.

A thoroughly classic pedigree, inbred to Lord Clifden at 3rd and 4th rems. and to Stockwell at the 4th. At the time of mating Isinglass would not have been a proven classic sire, but Mayonnaise, through a sister to Corrie, would have been known as the granddam of Bruce 1879 GP. Later she was to become ancestress of Homeward Bound 1961 O.

Glasalt, a first foal, won 2 races.

She had altogether 9 foals of which 6 won 24 races and £17,300; of those 6, four were by sons of St. Simon and Glacier by the great horse himself.

Glacier was inbred to Galopin at the 2nd and 4th removes, Canyon at the 3rd and 5th.

GLASALT 1898

Mated with: – ST. SIMON

Produce: – 1907 GLACIER
won 2 small races
bred 7 wnrs. of
23 races – £40,000

TOBOGGAN 1925
O, Hurry On
won 7 races

BOBSLEIGH 1932
Gainsborough
inbred St. Simon
3 & 4 rem. won
3 mid. dist. races
in 3 seasons.
Sired Moonlight
Run (USA) and
the dams of
*†Mossborough
*Sovereign Path
*Kalydon

HYDROPLANE II
1938 Hyperion
exp. USA 1941
did not win

***CITATION 1945**
Bull Lea
USA Champion
won Ken.D, Bel.,
Preakness etc.

Mated with: – CHAUCER (St. Simon's son)

Produce: – 1913 CANYON 1.
won 3 other races,
bred 7 wnrs. of 33
races – £51,600

***† COLORADO 1923 2, E, Phalaris** inbred St. Simon 3 & 4 rem. and Hampton at 4th.
Colorado won altogether 9 races and £30,300; he was 3rd in the Derby and later won over 12f. This is the classic formula referred to earlier on – sire Phalaris, dam by Chaucer. Colorado unfortunately died after 2 seasons at stud during which he got Loaningdale E, *Felicitation GC, Colorado Kid, Scarlet Tiger and Furrokh Siyar. His daughters bred Commotion O and *Clarion.

and his brother

***CAERLEON 1927**
E, Phalaris
sired the dam of Zabara 1, the 2nd d. of *St. Paddy D, L, E, and the 3rd d. of Pourparler 1, and Fleet 1.

ERYHOLME 1898

The origins of **ERYHOLME** 1898. By **HAZLEHATCH** 1885 who won 6 races from 5f to 1 mile and £4,200, was not very distinguished; by 1900, when exp. to Germany, his fee was 18 guineas. By Hermit from an unexciting bottom line, he was, however, inbred closely to Newminster at 2nd and 3rd removes.

Ex AYRSMOSS 1894	by Ayrshire, did not win.
Ex RATTLEWINGS	by Galopin, sis. to Galliard 2.
Ex MAVIS	by Macaroni.
Ex MERLETTE	by The Baron, and 5 gens. back to the foundation mare Fanny by the first Derby winner, Diomed. Family no. 13.

Eryholme was the first and only foal of her dam, she did not run. She is inbred at the 3rd and 4th removes to both Newminster and Galopin.

She had 12 foals to live, three of them won 15 not very important races.

ERYHOLME'S MATINGS:

***Love Wisely** 1893 GC (at 3y), a grand stayer by the useless racehorse but first-class sire, Wisdom. His dam was also stayer-bred by the long-range Philammon.

Wavelet's Pride 1897, another stayer, by Isonomy's unimportant brother, Fernandez. Wavelet's Pride lived to be 24 and became a great sire of jumpers. His dam, the unfashionably bred **Wavelet**, by Paul Jones, won 23 races out of 75 runs under both Rules.

Barcadaile 1901, to whom can be awarded no other distinction than that of being Flying Home's sire, was more nobly bred, being by St. Simon's son, St. Frusquin from a Barcaldine mare. This put Galopin 3 times into Flying Home's pedigree, twice at the 4th and once at the 5th remove.

ERYHOLME 1898

Mated with: – LOVE WISELY

Produce: – 1905 ANCHORA
won 8 small races,
bred 5 wnrs. of
£6,200.

SCAPA FLOW
1914
Chaucer
won 3 small races, bred 8 wnrs. of 63 races and £86,000.

***†PHAROS 1920 Ch. Phalaris**
Whereas his brother Fairway won at up to 18f, Pharos's best distance was 10f. Although he won 14 out of the 30 races he ran in, none of them, with the exception of the Champion Stakes which he won and the Derby, in which he was 2nd, were of much consequence. He stood both in England and France, becoming champion sire once in the former and twice in the latter. Nearco was of course his greatest son and next, I suggest, comes *†Pharis II FD, GP, who sired four FD wnrs. and *Pardal; nevertheless there were 9 others of considerable note including *†Cameronian 2, D, Ch.

and his brother inbred St.Simon 3 & 4 rems.

***†FAIRWAY 1925 L,E,Ch. (twice) Phalaris**
Fairway won 8 other notable races and he might have won the Derby had it not been for his immense popularity which caused the crowd to mob him to the extent that he was badly upset. He sired the winners of 9 classics and got 8 important stallions including *†Blue Peter 2, D, E, *†Honeyway Ch. and *†Fair Trial.

and their sis.

FAIRISLE 1927 1, Phalaris
bred 1 good wnr.

ROTHESAY BAY
1916 Bayardo
won 4 races,
bred 8 wnrs.

BRODICK BAY
1922
Swynford
won once
bred 2 wnrs.

MIRACLE 1929 E, Manna
failure at stud.

PLADDA
1926
Phalaris
won 4 races
bred 4 wnrs.

PLASSY 1932 Bosworth
won Coronation Cup and Jockey Club Stakes. Inbred Bay Ronald and Cyllene at 4th rem. Exp. France 1936. Sired *† Vandale and Nepenthe.

WAVELET'S PRIDE
|
1910 CELONE
did not win
bred 4 wnrs.
|
ATHLONE 1915
Orby
Sired the d. of
Solenoid

BARCADAILE
|
1918 FLYING HOME
did not win
bred 6 wnrs.
£13,300

BLANDING 1931
Blandford
sprinter, sired dam
of Arctic Storm Ch.

ANNE HATHAWAY
1926 Stratford
won 5 races
|
LAVINIA 1937
Bosworth
won twice
|
WILD SUCCESS
1949 Niccolo
del'Arca
won 1½ races
£2200
|
PLAZA 1958
Persian Gulf
won 2 good races
|
INTERMEZZO
1966 L, Hornbeam
exp. Japan 1970

SANTA BRIGIDA 1898

The origins of **SANTA BRIGIDA** 1898. By ST. SIMON.

Ex BRIDGET 1888	by the stayer *MASTER KILDARE 1875. Bridget did not win. Her own brother, ***†MELTON** 1882 D, L, had distinguished himself twelve years previously, and this mating with St. Simon (the first and only time) must indeed have raised high hopes.
Ex VIOLET MELROSE	by The Scottish Chief
Ex VIOLET	by Thormanby. Founded 2 other lines.
Ex WOODBINE	by Stockwell see p.v. Family no. 8.

Santa Brigida won races to the value of £4,700.

She had 10 foals eight of which won 31½ races and £20,600.

All her matings were outcrosses except that Ayrshire brought to Brig Of Ayr two strains of Galopin at the 3rd remove.

SANTA BRIGIDA 1898

The fourth of the St. Simon mares

Mated with:– / Produce:–

LOVE WISELY

- **1903 BRIDGE OF CANNY** first foal. won 15 races and £14.500. **Exp. Argentina** 1911 having sired Cantilever (Jockey Club Stakes), mat. grandsire of *Felicitation GC.

ISINGLASS

- 1904 ST. CLAIRE II won once
  - LADY CICERO 1913 Cicero exp. France 1925
    - REGAL ROMAN 1921 Roi Herode Exp. USA 1923
      - ALCIBAIADES 1927 Supremus won Ken. O.
        - ***MENOW (USA) 1935 Pharamond II** Sired *†Tom Fool and the 2nd d. of *Baldric II 2, Ch.
        - SALAMINIA Man O'War
          - ATHENIA 1943 Pharamond II
            - ATTICA 1953 Mr. Trouble
              - **SIR IVOR 1965 2, D, Ch. (Wash. Int.) Sir Gaylord** Imp. UK 1966 exp. USA 1970

ISINGLASS

- sis. 1905 BRIDGE OF SIGHS did not win bred 4 wnrs. of 21 races – £13,000
  - **LIGHT BRIGADE 1910 Picton** Inbred Isonomy 3 rem. won 16 races and £11,400. Exp. USA 1916. Sired d. of †Discovery.

CYLLENE

- **1906 BRIDGE OF EARN** A stayer, sired Bulger and the dam of Cresta Run 1.

AYRSHIRE

- 1907 BRIG OF AYR won twice, bred 5 wnrs.
  - GARPAL 1920 Phalaris won once at 2y. bred 3 wnrs. of 22½ races
    - **BROWN BETTY 1930 1, exp. USA '33. Friar Marcus** Inbred Cyllene at 3 & 4 rems. and to St.Simon at 4th.
  - and her sis. COILA 1922 Phalaris won once at 2y.
    - CURL PAPER 1931 Papyrus did not win
      - PERMAVON 1936 Stratford won once at 2y.
        - MARCELETTE 1942 William Of Valence did not win bred 3 wnrs.
          - **MARCH PAST 1950 Petition** ran 23 times and won 10 races, his best performance being the 7f Greenham at Newbury, which could be looked upon as a mild classic trial; he was, however, hopelessly outclassed in the 2000 gns. Aged 4 he won the 6f Wokingham at Ascot carrying equal top weight, 9 stone. Looking at his pedigree it would seem that his speed came from Petition, with his Lady Josephine background, rather than from his dam's side. He died in 1976 and up to time of writing has got over 450 wnrs on the flat and quite a few over obstacles. see p. 158.

SPEARMINT

1910 SPEAN BRIDGE
won 3 small races
bred 5 winners.

BRIDGE OF ALLAN
1928 Phalaris
won 1 small race

MIDDAY SUN 1934
D, Solario
to stud 1939, perhaps
unlucky because of
the war. Exp. N.Z. 1950
died 1954

NEIL GOW

1916 SANTA CRUZ
won 4 races exp. Fr.'26
bred 7 wnrs. of 18½ races
£12,800

DRIFT fld. Fr.
1926
Swynford
won 4 races
& £3309
bred 5 wnrs.
inbred St. Simon
3 & 4 rems.

TIDEWAY 1933 1,
Fairway
Inbred St. Simon
4 rem.

GULF STREAM
1943 E, Hyperion
Inbred Chaucer
3 & 4 rems.
Exp. Argentina
1947.

***HELIOPOLIS**
1936 Hyperion
won 5 big races,
Champion sire
USA

& his brother

SEVEN SEAS
1941 (exp. USA
'51) Hyperion
sired 2nd dam
of Roseliere FO.

& sister

SUN STREAM
1942 1, O,
bred 4 wnrs.

These three inbred to St. Simon at 4th rem.

GOODY TWO SHOES 1899

The origins of **GOODY TWO SHOES** 1899. By ISINGLASS.

Ex SANDAL 1885	by ***KISBER** 1873 D, GP, ½bro. to Wenlock, bred in Hungary, his dam Mineral having been sent there in 1871. He was a success there at stud, his daughters being connected with several French classic winners and good stallions.
Ex SHOESTRING	by Knight Of The Garter.
Ex BOWSTRING	by Tom Bowline.
Ex a bay mare 1861	by Toxophilite, and back 7 gens. to the foundation mare Miss West 1777 by Matchem. Family no. 5.

Goody Two Shoes won one race.

She had 7 foals, all by good stallions; her pedigree was free of St. Simon, a matter that was soon put right by her matings:- ***†DESMOND** 1896 by St. Simon from the Oaks winner **L'Abbesse De Jouarre** by Trappist by Hermit. A good winning 2yo, Desmond turned naughty later and would not try, a tendency which he did not pass on to his offspring. Although, besides Charles O'Malley, he sired no less than 11 distinguished horses, including the winners of 4 classics and both winner and disqualified second in the disastrous Derby of 1913 – (suffragette killed bringing down the King's horse Anmer (unhurt) at Tattenham corner, Craganour disqualified in favour of Aboyeur; neither of them much good) – none of his male get except Charles O'Malley and Farasi have made impact of top-level importance. His daughters however, produced *†Grand Parade D, Bettina 1, *†Sir Cosmo, *Obliterate, Dominion, Collaborator and Spike Island.

SIMON SQUARE 1904 by St. Simon from a mare by Kendal (by Bend Or), won 10 races as a 2yo, beyond which it seems that Simon's Shoes represents his only claim to fame, and a pretty good one it is.

GOODY TWO SHOES 1899

Mated with: – DESMOND *(by St. Simon)*

Produce: –

1907 *† CHARLES O'MALLEY
A stayer, won Gold Vase. Sired Charlebelle O, and the dams of *†Blenheim D, *†King Salmon E, †His Grace and *Bacteriophage.

Mated with: – SIMON SQUARE *(also by St. Simon)*

Produce: –

1914 SIMON'S SHOES
did not win, bred 8 wnrs. of 21½ races – £10,100

- CARPET SLIPPER 1930 Phalaris
  won 1 mod. race
  inbred St. Simon 3 & 4 rems.
  bred 5 wnrs. of 16 races – £13,400
  - HER SLIPPER 1936 Tetratema
    did not win
    exp. France 1950
    - VALI 1956 Sunny Boy
      - ***VAL DE LOIR 1959 FD**
        Vieux Manoir
        died a. 10
        sired Tennyson GP
        Val De Lorne FD
      - KALILA 1961 Beau Prince II
        - **ROI LEAR 1970 FD,**
          Reform
      - **VALORIS 1963 O,**
        Tiziano
        *TIZIANO (It.L) by Sicambre from a dau. of Tofanella (d. of Tenerani)
  - **GODIVA 1937 1,O.** **& her sis.**
    Hyperion
    Inbred St. Simon 4 rem.
  - COVENTRY BELLE 1938 Hyperion
    won once
    bred 4 wnrs.
    - MISS COVENTRY 1943 Mieuxce
      did not win
      - COUNTRY HOUSE 1955 Vieux Manoir
        won 3 races
        bred 7 wnrs.
        - ***REFORM 1964 Ch.**
          Pall Mall
          has sired so far Roi Lear (above) and Polygamy O. Reform's Sire, *PALL MALL 1955 2 and six other races, was by Palestine from Malapert by Portlaw and inbred to The Tetrarch at the 4th rem.
  - ***WINDSOR SLIPPER 1939 (Ir.2,D,L)**
    Windsor Lad
    never beaten, sired *†Solar Slipper 1945 Ch. (exp. USA 1957)
- DALMARY 1931 Blandford
  won 1 good race
  bred 5 wnrs.
  - JENNYDANG 1938 Colombo
    won 2 small races
    bred 5 wnrs.
    - ERICA FRAGRANS 1946 Big Game
      won once
      - PHOENISSA 1951 The Phoenix
        won twice,
        bred 5 wnrs.
        - **LORENZACCIO 1965 Ch.**
          Klairon
          *†KLAIRON 1952 (F2) Imp. UK '62 By Clarion III from a Kantar mare; inbred to Gay Crusader at the 4th rem.; Sired also Monade O.

SCEPTRE 1899 2, 1, O, L, Ch.

The origins of **SCEPTRE** 1899 2, 1, O, L, Ch. By Persimmon from Ornament q.v. p. 7.

It is hard to imagine what more there was for this prodigy, Sceptre, to have won; but apparently she competed also in the Derby, Grand Prix and Lincolnshire handicap. She won the Hardwicke and Jockey Club Stakes but got beaten in the Coronation and St. James' Palace Stakes. She is the product of about the most classic breeding anyone could invent, and is an outcross for 4 generations. Her dam, Ornament, visited Persimmon on three occasions resulting in 1899 Sceptre, 1900 barren (Ornament was barren 5 times altogether) and 1903 a sister called Crown Gem who bred only small winners.

Sceptre had 8 foals to live, was barren 9 seasons and died in 1926. In the matings on the next page, Maid Of The Mist is inbred to Bend Or in the 3rd generation. Later appears Cyllene's son *†**CICERO** 1902 D, superbly bred out of Gas by Ayrshire. He made himself famous at stud by getting, from another Persimmon mare, *†Friar Marcus, one of the greatest sires of broodmares.

Curia and her brother Grosvenor were inbred to Bend Or at the 3rd and 4th and to Hampton at the 4th removes.

SCEPTRE 1899 2, 1, O, L, Ch.

Mated with:– CYLLENE

Produce:– 1906 MAID OF THE MIST
won 3 races
bred 6 wnrs.
of £23,000

HAMOAZE
1911
Torpoint
won twice
bred 5 wnrs.
of 25 races
£37,400

- ***†BUCHAN** and bro.
1916 E (twice)
Ch. Sunstar
very versatile,
won Don. Cup.
disq. GC & a
close 2nd in
both 2000 and
Derby.
Champion sire
once. Sired
Short Story O,
Booklaw L and
the dams of Sun
Castle L, Airborne
D, L, *†Fair Copy
and Pay Up 2.
- **SALTASH**
1920 E,
Sunstar
exp. Australia
1924
- **ST. GERMANS**
1921
Swynford
Won Don. &
Coronation
Cups, Sired
St. Brideaux.

SUNNY JANE & her bros.
1914 O,
Sunstar
Inbred Bend Or
4 rem. exp. USA
1926

MISS CAVENDISH
1919 Chaucer
bred 5 wnrs.
£21,100

NEVER CROSS
1924 Gay
Crusader
won once

GERRARD'S CROSS
1930 Teddy
won once

MISS MINX
1936 Mr. Jinks
won once at 2y
bred 7 wnrs.

WHEEDLER
1943 Umidwar
won 5 races.
bred 10 wnrs.

***COUNSEL**
1952
Court Martial
won 10 races
av 8f. Sired
Privy Councillor 2.

SKY ROCKET and
1915 Sunstar
see p.20, App. D.
Inbred Bend Or
Sky Rocket was
a good 2y who
deteriorated.

***CRAIG AN**
ERAN
1918 Sunstar
2, E (3y)
Inbred Bend Or,
ran a close 2nd
in the unhealthy
Humorist's Derby,
some people
blamed the jockey;
anyway it was a
pity he did not
win. He sired 4
classic winners,
the best of which
has proved to be
*†Admiral Drake
GP.

JURA 1921
Gainsborough
won 3 good
races

GLENABATRICK
1927 Captain
Cuttle
did not win

TIBERIUS
1931 GC,
Foxlaw
Inbred Cyllene
4 rem. Also
won Gdwd.
Cup. No use
at stud.

ISINGLASS

1908 CORONATION IV
Visited Italy 1911,
returned 1920

- THEOTOCOPULA
  1914 Thrush
  - TURLETTA 1921
    Sir Archibald
    - TORRIGIANA
      Ortello
      - TRESA 1946
        Bozzetto
        - **TAMANAR 1955**
          FD, Sunny Boy
          Inbred Gainsborough
          and Teddy 4 rem.
- TERRA D'OMBRA
  1918
  Sir Archibald 1905
  (exp. Fr. '16)
  - CREPUSCULE
    1925
    Galloper Light
    - STELLA POLARIS
      1937 Papyrus
      - **NORTHERN LIGHT**
        1950 GP,
        Prince Bio
      - POLAIRE 1947
        Le Volcan
        - RELANCE III
          1952 Relic
          - ***MATCH III**
            1958 GVI, imp.
            UK '62 (FL,
            Wash. Int.)
            Tantieme
          - & his bro.
            RELIANCE II
            1962 FD, GP
            Tantieme
            Imp. UK 1966
          - ***RELKO**
            1960 D,
            Tanerko

- KEENER 1930
  Coronach
  did not win
  - FRENCH KIN
    1938
    Brantome
    won once
    - **MITRAILLEUSE**
      1944 Mieuxce
      won 3 good races,
      bred 11 wnrs.UK
      - MITRAILLE
        1953
        Big Game
        did not win
        bred 5 wnrs.
        - FUSIL 1961
          Fidalgo
          won once
          bred 4 good
          wnrs. so far
          - **FULL DRESS II**
            1961, 1, exp USA
            '71 '76
            Shantung
      - TRIAL BY FIRE
        1958
        Court Martial
        - SINGE 1972
          Tudor Masie
          - ONE IN A MILLION 1
            1976 Rarity

CARBINE

1909 QUEEN
CARBINE
won 1 small
race

- QUIRE 1918
  Fairy King
  did not win
  bred 4 wnrs.
  - ART PAPER
    1933
    Artist's Proof
    won twice,
    bred 3 good
    wnrs.
    - ***†PETITION**
      1944 E, Fair
      Trial
      Champion sire
      once, his chef
      d'oeuvre being
      the filly Petite
      Etoile 1, O, Ch.
      (from a Bois
      Roussel mare).
      He also got the
      prolific March
      Past (see p.60)
      and the dam
      of Night Off 1,
      see also p.159.

CICERO

- 1912 CURIA & her bro
  did not win
  bred 7 wnrs.
  - ACQUIT
    1928
    Hurry On
    won ½ race
    - **FLYON**
      1935 GC
      Flamingo
      no good at
      stud inbred
      Persimmon
      4 rem. His
      sire
      FLAMINGO
      2, by Flam-
      boyant (a
      stayer) from
      a White Eagle
      mare, was
      inbred to
      St.Simon &
      Galopin.
- **GROSVENOR**
  1913
  won 1 decent
  race, inbred
  Bend Or 3 &
  4 Hampton 4
  rems. Sired 3rd
  dam of Souepi
  GC & got many
  wnrs. of small
  races, exp.
  Australia '26
  and did no
  better there.

GLENESKY

1917 QUEEN
EMPRESS
did not win

- GRAND
  PRINCESS
  1922
  Grand Parade
  did not win
  - QUEEN OF
    SCOTS 1930
    Dark Legend
    did not win
    bred 7 wnrs.
    - QUEEN OF
      BAGDAD
      1937 Bahram
      did not win
      bred 6 wnrs.
      - QUEEN OF
        BASRAH
        1946 Fair
        Trial won
        once, bred
        2 wnrs. exp.
        Fr '55.
        - **TABOUN**
          1956 2,
          Tabriz
          inbred
          Blandford
          at 4th rem.

VALVE 1900

The origins of **VALVE** 1900. By †**VELASQUEZ** 1894 E, Ch. (twice), strongly bred by *†Donovan D, L, from Vista by Macaroni, the dam of Bona Vista 2, and Sir Visto D, L. He sired also the dams of Vaucluse 1, By Jingo GC, Volta and Nouvel An.

Ex GAS 1892	by Ayrshire, did not win, dam of Cicero 1902 D.
Ex ILLUMINATA	see p. 39. Family no. 1.

Valve ran 24 times and won 8 races, she is inbred to Galopin at the 3rd and 4th removes.

She had 12 foals by eleven different stallions, between them they won 20 races and £15,300. Verve was the product of her only mating with Velocity and was inbred to Hampton at the 3rd and 4th removes.

*†**DARK RONALD** 1905 broke down as a 3yo but next year won the Hunt Cup (car.7.2). He stood in England for only 3 years and was sent to Germany in 1913. During those 3 years, in addition to Vaucluse (over), he sired the mighty Son In Law 1911, responsible for so many fine stayers. He also got *†Dark Legend 1914, sire in France of the winners of 5 classics, but perhaps his most important produce was that grand mare Popingaol 1913 (p. 40). To this day in Germany there are few pedigrees without Dark Ronald in their backgrounds. He was sired by Bay Ronald from a mare by the GP winner, Thurio; so Vaucluse, like her half-sister Verve, is inbred to Hampton to the same extent.

VALVE 1900

Mated with: –

Produce: –

VELOCITY

1909 VERVE
did not win
bred 6 wnrs. of
25 races –
£10,100

RHONA 1920 (sis)
Hurry On

DOUBLURE
1933
Comedy King
bred 4 wnrs.

KING'S CROSS
1940 exp. USA
'57 King Salmon
bred 6 wnrs.

BLUE CROSS 1946
Blue Peter

GRAND CROSS
1952 Grandmaster

HARDIEMMA
1969
Hardicanute

**SHIRLEY HEIGHTS
1975 D, IRD,
Mill Reef**

**†KING'S BENCH 1949
Court Martial**
Inbred Hurry On
3 & 4 & Son In Law
4 rem. A very good
2yo, won St. James'
Palace Stks at 3, exp
Jap '68, Sired dam
of Prince Royal II AT.

CRADLE SONG
1922 Hurry On
bred 10 wnrs.

PAINTER'S SONG
1935 Gainsborough

BRAY MELODY
1942 exp. USA
'54, Coup de Lyon.
bred 10 wnrs.

**HAPPY LAUGHTER
1950 1 Royal
Charger**
Inbred Gainsborough
3 & 4 rems.

VERVELLE 1923
Buchan
bred 6 wnrs.

EMPIRE CRUSADE
1928 Orpheus
exp. Fr. 1933

MACREUSE 1938
Massine (Ap.A)

QUEMANEUSE
1943
Quai d'Orsay

NODULEUSE
1954 Nosca

**PRINCE REGENT
1966 IRD
Rt. Royal V.**
Imp. UK '69

APULIA 1931
Apelle
won 7 races
bred 6 wnrs.
exp. Fr. 1936

**AQUINO II
1948 GC,
Tornado**
won also Don. Cup
To stud in Poland.
His sire *†TORNADO
1939 won the Prix
Lupin & 4 other races.
He was by Tourbillon
from a Swynford mare;
he sired also Tahiti FO,
*†Fontenay and the
dam of *Misti.

DARK RONALD

**1912 VAUCLUSE 1
(her only win)**
bred 5 wnrs. of £8,800

LAKE LEMAN 1920
Tracery
won twice

EDGELAW 1928
Ellangowan
won twice
bred 2 wnrs. of 15 races

**HIPPIUS 1937
Ch. (twice)**
exp. Brazil 1941, died
en route.

OUR LASSIE 1900 O

The origins of **OUR LASSIE** 1900 O. by AYRSHIRE p.35, App. D.

Ex YOURS 1894	by MELTON p.19, 55, App. D. Bred in Italy, imp. UK 1897, Yours did not win. Dam later of **Your Majesty** 1905 L, E, (exp. Arg. '11) by Persimmon.
Ex YOUR GRACE 1886	by Galliard. Your Grace exp. It. '88, returned '96. 2nd dam of Or Du Rhin FD.
Ex GRAND DUCHESS 1871	by Lozenge. Had 13 foals four of which founded classic branches.
Ex LADY-LIKE 1858	by Newminster. Dam of †**Rosebery** 1872 who won both Cambridgeshire and Cesarewitch as a 4yo, those races being his objective. He was far better than a handicapper.

Our Lassie was inbred to Galopin at the 3rd and 4th removes.

She only bred 2 winners of 5 races from her 8 foals.

OUR LASSIE'S MATINGS:

Mated with ***†ORME** 1889 E (twice), Ch. was by Ormonde out of St. Simon's sister Angelica. He was champion sire on one occasion; his most important son was *†Flying Fox 1896 2, D, L, E, the progenitor of Teddy; 9 years later he got the Derby winner Orby and Witch Elm 1. Sweet Lassie (above) has 3 strains of Galopin, 3rd, 4th and 5th removes.

OUR LASSIE 1900 O.

Mated with: –
Produce: –

- Mated with ORME
  - **Produce:** 1906 SWEET LASSIE first foal did not win bred 3 wnrs. of 10½ races
    - **HAPPY WARRIOR 1911 Sundridge** Sired the dam of *†Panorama
    - and his brother **SUNFIRE 1912 Sundridge** Sired 4th dam of Belle Ferroniere FO
- Mated with SUNDRIDGE
  - **Produce:** 1912 LADY BRILLIANT won once, bred 3 wnrs.
    - BLACK GEM 1918 Black Jester did not win bred 7 wnrs.
      - **BULGER 1923 Bridge of Earn** won 8 races of up to 10f. Sired 3rd d. of Privy Councillor 2, and Weaver's Hall IRD
      - and his sis. (Inbred Cyllene 2 & 4 rems.) BRIG O'DEE 1926 Bridge of Earn won twice, bred 4 wnrs.
        - INDISCRETION 1932 exp. Fr. '39 Hurry On did not win bred 6 wnrs.
          - **IMPRUDENCE 1944 1, O, Canot (F1)**
    - and her sister BLACK RAY 1919 Black Jester won 1 small race bred 10 wnrs.
      - INFRA RED 1936 Ethnarch won 2 good races bred 4 wnrs.
        - RED RAY 1947 (died aged 6) Hyperion did not win exp. USA 1949
          - VIRGINIA WATER 1953 Count Fleet
            - MILAN MILL 1962 Princequillo
              - **MILL REEF 1968 (foaled USA) D, GVI, E, AT, imp. '69. Never Bend** see p. 152
        - EXCELSA 1949 Owen Tudor
          - **WICHURAIANA** 1963 Worden
            - **WOLLOW 1973 2, E (on doping disq.*) Wolver Hollow**

** The French Trepan was tested with positive results.*

PRETTY POLLY 1900 1, O, L, Ch.

The origins of **PRETTY POLLY** 1900 1, O, L, Ch. (and of her sisters Adula 1902 and Miranda 1903). By GALLINULE see p. 13.

Ex ADMIRATION by Saraband, see p. 31.

Pretty Polly raced for four seasons winning 22 out of 24 races including, in addition to the classics, the Coronation and Jockey Club Cups. In her two defeats she was second. Family no. 14.

Of Pretty Polly's 10 foals, four were fillies and they *all* became classic ancestresses; they were each by a different sire, a fact which emphasises her great potency. ***†WILLIAM THE THIRD** 1898 GC was mentioned as the sire of Roseland, but requires some elaboration here. He was a great stayer, by St. Simon from Gravity by Wisdom (neither of them ever won) and, although he got 2 classic winners, it was his daughters who made him famous: between them they bred the winners of 7 classics. ***†POLYMELUS** 1902 Ch. is perhaps the most interesting of the progenitors in the Bend Or stirp; he is a very vital link, being the sire of Phalaris who now stands at the head of the lines of today's most prolific classic sires. As a racehorse Polymelus was talked of as a non-stayer, but it must be remembered that he ran 2nd in the Leger and also won over 1½m; nevertheless the av. winning distance of his progeny was 8.7f. Five times champion sire (beaten this century only by Hyperion), he got the winners of 11 classics. He was by Cyllene from Maid Marion by Hampton. Polly Flinders was consequently inbred to Isonomy at the 3rd and 4th removes. ***†SPEARMINT** 1903 D, GP, (see Sirenia & Concertina) by the tough stayer Carbine (33 races in the Antipodes inc. the Melbourne Cup carrying 10.5) from a mare by the GP winner, Minting, became one of the greatest sires of broodmares ever. His daughters bred the winners of 8 classics (Seaweed and Catnip see later). His son, Spion Kop, won the Derby. Spearmint was inbred to Stockwell at the 4th remove.

PRETTY POLLY 1900 1, O, L, Ch.

The greatest of Admiration's four great daughters

Mated with: –

Produce: –

- DESMOND
  - 1914 MOLLY DESMOND founded a family of her own see p. 74, 111, 113, 160, App. D.
- WILLIAM THE THIRD
  - 1915 DUTCH MARY did not win, bred 4 wnrs. of 8 races – £13,900
    - DUCCIA DI BUONINSEGNA 1920 Bridge of Earn won It.1. foaled Ireland, exp. Italy 1921
      - DELLEANA 1925 imp UK and visited It. and Fr. often. Clarrissimus. won both It 2 & 1.
        - DOSSA DOSSI 1930 Spike Island (It.1,O,) bred 4 wnrs.
          - DAGHERIOTIPA 1936 Manna (It. 1.) bred 6 wnrs.
            - DINARELLA 1947 Niccolo Dell 'Arca won once bred 1 other UK wnr.
              - ***PSIDIUM 1958 D, Pardal** Inbred Phalaris 4 rem. Exp. Argentina 1970. Sired Sodium IRD, L.
        - ***† DONATELLO II 1934 (It.D) fld Fr. imp. UK '38 Blenheim** won also 7 other important races in Italy, his only defeat was to have been 2nd in the GP. see p. 168
    - KOPJE 1925 Spion Kop exp. France finally 1933
      - **CAPPIELLO 1930 GP, (Lupin) Apelle**
    - SISTER ANNE 1926 exp. France 1937 Son In Law won once
      - PRUDENT POLLY 1944 Atout Maitre won once bred 4 UK wnrs.
        - **OSSIAN II 1952 Royal Charger** Sired dam of Pistol Packer FO.

POLYMELUS

1918 POLLY FLINDERS
won 1 race – £2,549
bred 6 wnrs. of 14 races – £15,200

ARABELLA
1926 Buchan
won 4 races
£10,500
bred 7 wnrs.

OVERMANTLE
1931 Apron
won twice

OVERTURE
1937 Dastur
did not win
bred 11 wnrs. of 31 races

FORECOURT
1943
Fair Trial
won once
bred 6 wnrs.

†SUPREME COURT 1948 GVI
Precipitation
Sired dams of Roll of Honour GP Random Shot GC Rheingold AT

FLIRTING 1945
Big Game
won twice,
bred 7 wnrs.

ARCTIC EXPLORER 1954 E, Arctic Prince
Inbred Solario 4 rem. Exp. Australia 1958. His sire †ARCTIC PRINCE 1948 D, by Prince Chevalier from a Nearco mare, exp. USA 1956 sired also the dams of Santa Claus D, IRD (Ir. 2) & Park Top GVI

ANNABEL
1932
Blandford
won 3 races
bred 4 wnrs.

BOROBELLA
1942
Bois Roussel
won once
bred 5 wnrs.

LIFE SENTENCE
1949
Court Martial
won 4 races
bred 7 wnrs.

ONLY FOR LIFE 1960 2,
Chanteur II
(see Ap. B under Pinza) inbred Blandford 4 rem. exp. Japan 1967

PARTICIPATION
1942
Precipitation
won twice
bred 8 wnrs.

NEUTRON
1948
Hyperion
won twice

***COURT HARWELL 1954 exp. Arg. '61**
Prince Chevalier
Sired Meadow Court 1962 IRD, GVI.

SPEARMINT

1924 BABY POLLY
did not win
bred 6 wnrs. of 18 races – £15,000

COLORADO KID 1929
Colorado
won the Hunt Cup, Jubilee and, perhaps surprisingly, the Doncaster Cup which was said that year to have been run at a crawl. He sired the dam of †Grandmaster and the 2nd dam of Hafiz II Ch.

RUSK 1935
Manna
won 1 small race

TURKISH BLOOD 1944
Turkhan
won twice
bred 5 wnrs.

***VIENNA 1957**
Aureole
exp. Fr. '67 then exp. Japan 1972 Vienna won 6 races inc. the Blue Riband Trial Stakes; he was a good horse but definitely sub-classic on the racecourse. He has earned his position in the Hampton dynasty by having sired *Vaguely Noble 1965 AT. see p. 173

SUNSHOT 1901

The origins of **SUNSHOT** 1901. by **CARBINE** see p. 43.

Ex STREAM OF GOLD 1895	by †**ST. ANGELO** 1889, 2nd in the 2000 gns., won the St. James' Palace Stakes; sired the dams of St. Astra FO and the important USA sire, North Star III. Stream of Gold ran 24 times, won once, exp. Germany 1906.
Ex GOLD STREAM	by Wisdom
Ex QUICK STREAM	by Trumpeter.
Ex QUICK MARCH	by Rataplan and 3 generations back to Nan Darrell 1844, whose daughters founded 3 classic branches of Family no. 19.

Sunshot won one small race.

She had 15 foals to live. The two successful Bridge Of Earn (p.60) daughters are inbred to Galopin at the 4th remove.

SUNSHOT 1901

Mated with: –

Produce: –

BRIDGE OF EARN

1914 SUNBRIDGE
won 2 small races
bred 6 wnrs. of
£13,600

RESPLENDENT
1923 (Ir. 1, O)
By George
bred 4 wnrs.

***†WINDSOR LAD**
1931 D, L, E,
Blandford
won 10 out of 13
races including the
Coronation Cup.
His time, winning
the Leger, equalled
the record. He
died after only 3
seasons at stud,
having got *Windsor
Slipper and the
unbeaten Phase,
dam of Neasham
Belle O and
Narrator Ch.

SOL SPERANZA
1924 (Ir. 1, O)
Ballyferis
bred 4 wnrs.

†SOLFERINO
1940 (Ir.L)
Fairway
inbred Cyllene
4 rem. Sired
*Solonaway
and the dam of
Oroso AT.

QUEEN SCOTIA
1926
Tetratema
won twice

SCOTIA'S GLEN
1935
Beresford
won twice
bred 3 wnrs.

GLEN LINE
1942
Blue Peter
did not win,
bred 4 good wnrs.

***KING OF THE**
TUDORS 1950
E, Tudor Minstrel
exp. USA 1959
Sired Henry The
Seventh E.

OUR BABU, 1952
2, My Babu
No great success
at stud exp. USA
'55 and Japan '67
His sire *† MY
BABU 1945 2, won
altogether 11½ good
races. By Djebel from
Perfume II (d. of
Sayani) by Badruddin.
He sired also *Primera
(exp. Japan) My Babu
was exp. USA 1956.

FLYING ORB

1917 DONATI'S
COMET
won 3 races in Ire-
land bred 5 wnrs.

MRS SWANSONG
1940 Sir Walter
Raleigh
won 3 races,
bred 4 wnrs.

CYGNET 1950
Caracalla II

***WOLVER**
HOLLOW
1964 E,
Sovereign Path
Sired Wollow 2,E.
His sire: *SOVEREIGN
PATH 1956, av.
winning dist. 7f, by
Grey Sovereign from
a Bobsleigh mare,
sired also Humble
Duty 1, and Supreme
Sovereign.

BRIDGE OF EARN

1919 KNOCKANEY
BRIDGE (sis. to
Sunbridge) won
1 small race, exp.
USA 1924

BLOOD ROOT
1932
Blue Larkspur
won 8 races

BE FAITHFUL
1942 Bimelech
won 14 races

LALUN 1952
Djeddah
(Ken. O)

***NEVER BEND**
1960 Nasrullah
Sired Mill Reef
D, GVI, E, AT.
see p. 152

ALTOVISCAR 1902

The origins of **ALTOVISCAR** 1902. By the stayer DONOVAN (p. 5).

Ex NAVARETTA 1893	by **KILWARLIN** 1884 L, a tough stayer by Arbitrator from the dam of Bendigo by Lord Gough. He sired also Ogden, a big winner in America, and the enduring Kroonstadt who ran 91 times in 8 seasons and won 22 races but whose chief triumph was siring the mare Querquidilla, dam of the immortal Brown Jack. Navaretta did not run, exp. France 1907.
Ex PAMPELUNA	by The Palmer.
Ex CATALONIA	by Lord Clifden.
Ex TARAGONA	by Orlando, a granddaughter of the Oaks winner INDUSTRY 1835, founder of 4 classic lines. Family no. 2.

Altoviscar won once at 2y.

She had 11 foals, 5 of which won 14 quite ordinary races.

ALTOVISCAR'S MATINGS:

Gallinule by this time was established as a great sire and the production of Alope was a nice outcross of the Isonomy and Galopin lines. Then –

*†**BACHELOR'S DOUBLE** 1906, won the Hunt Cup and middle-distance handicaps to the tune of over £10,000. His breeding, by a successful sire of the line of Bend Or, from a granddaughter of Isonomy and inbred withal to Hermit at the 3rd and 4th removes reads well, to say the least; examination of it, however, reveals it as little short of a fairy story. His sire, Tredennis (by the champion Kendal), raced for 3 seasons and never won a race, went to Ireland and started stud duties covering mostly ½bred mares; however, a few thoroughbred mares visited him and he began to get winners. So, in course of time, he got better mares and by the time he died aged 28 he had got the winners of 480 races. Bachelor's Double's dam was Lady Bawn, a non-winning twin; her sire, Le Noir (by Isonomy), never ran, was sent to Belgium as a 2yo and returned to Ireland and oblivion later. Although Bachelor's Double sired the winners of 3 classics, his fame is derived from his daughters. They bred: Pillion 1, Call Boy D, *†Fiterari GP, Duplex FD, Lovely Rosa O, *†Precipitation GC, Scottish Union L and Bellini (It.D,L).

ALTOVISCAR 1902

Mated with:– GALLINULE

Produce:– 1909 ALOPE
2nd in 2000 gns.
won 4 races
bred 5 wnrs.
of £16,500

***FOXLAW 1922**
GC, Son In Law
Inbred Donovan
3 and 4 rems.
Sired *†Foxhunter
GC and Tiberius
GC.

and his sister

ALOE 1926
Son In Law
did not win,
bred 5 wnrs.

FEOLA 1933
Friar Marcus
founded family
of her own.
See later.

SWEET ALOE 1935
Cameronian
did not win.

CHENILLE 1940
King Salmon
did not win
bred 6 wnrs.

LIGHTNING
1950
Hyperion
Won once at 2y
bred 6 wnrs.

***PARTHIA**
1956 D,
Persian Gulf
exp. Japan 1968
Sired Sleeping
Partner O.

***ALCIDE 1955**
L, GVI,
Alycidon
Died 1973, no
classic wnr. or
dam of wnr. so
far.

Mated with:– BACHELOR'S DOUBLE

Produce:– 1913 COMEDIENNE
won 5 small races, bred
3 wnrs. of 13 races and
£18,600 exp. France '32

COMEDY STAR
1921 Sunstar
did not win
bred 7 wnrs.

SHOW GIRL
1926
Son In Law
won 1½ races

DANCING TIME
1938 1, Colombo
Inbred Sunstar
3 and 4 rems.
Her sire *COLOMBO
1931 2, won 9 races
in all. By Manna from
a Chaucer mare;
inbred St. Simon at
4th rem.; Sired also
Happy Knight 2
and the dam of
*Tenareze.

COMEDY KING
1923
Son In Law
Inbred Donovan
3 and 4 rems.
won 7 good mid-
dist. races. Sired
2nd d. of †King's
Bench

CALL BOY 1924
D, Hurry On
beaten sht.hd.in
2000. Sired 2nd
d. of Democratic
and 3rd d. of Sweet
Solera 1, O.

GONDOLETTE 1902

The origins of **GONDOLETTE** 1902. By †**LOVED ONE** 1883, by the Hunt Cup and Cambridgeshire winner ***See-Saw** out of the dual Guineas winner Pilgrimage (p.37, 167, 168), was a moderate sprinting handicapper. He was 18 when mated to the equally ancient Dongola and Gondolette was her twelfth living foal, he had previously sired *Dinna Forget and the magnificent Doris, the selling-plater who produced 10 winners including *†Sunstar 2, D, and Princess 1, O.

Ex DONGOLA 1883	did not win had 13 other foals. Very well bred by *Doncaster D, GC.
Ex DOURANEE	by Rosicrucian.
Ex FENELLA 1869	by Cambuscan. Fenella was imp. from France 1870.
Ex LA FAVORITE	by Monarque, going back to Coral 1816, exp. from UK to France. Family no. 6.

Gondolette won one small race.

She had 12 foals and bred 10 winners of 24½ races value £32,700. All her 8 fillies eventually bore winners.

GONDOLETTE'S MATINGS:

White Eagle (p. 35) with his 2 strains of Isonomy and one of Galopin could be said to combine well with Gondolette whose pedigree is free of those two potent ingredients. Dolabella and Let Fly are thus inbred to Isonomy at the 3rd and 4th rems.

†**MINORU** 1906 2, D, (exp. Russia '73) was by Cyllene from a **Friar's Balsam** (by Hermit) mare; his g'dam was by Galopin. Serenissima was inbred to Hermit at the 3rd and 4th rems. Minoru was sent to Russia in 1913 but not before he had got the prolific Tillywhim (see later). The **Swynford** (p.36, 38, 168, App. D.) mating meant that Ferry and Sansovino were inbred to The Palmer at the 4th remove, where there was also his brother, Rosicrucian. *†**STEDFAST** 1908 Ch. won also the Coronation Cup, Hardwicke Stakes and Jockey Club Stakes and was 2nd to Sunstar in both 2000 gns. and Derby. He was by Chaucer from a mare by the 2000 gns. and Eclipse winner Surefoot. Stedfast's efforts at stud were confined (in terms of the classics) to siring Brownhylda O, whose son, Firdaussi, won the Leger.

GONDOLETTE 1902

Mated with:–

GALLINULE

Produce:–

1910 GREAT SPORT
Inbred Stockwell
4 rem. 3rd in Derby,
won 4 mod races.
Sired the mare
Bunworry, dam of
*† Bozetto and
important in Italy
as the 3rd d. of
Boticelli GC and
Bonnard (Don.Cup).

WHITE EAGLE

1911 DOLABELLA
won 1 small race
bred 7 wnrs. of
27 races –
£22,500

and her bro.

1912 LET FLY Ch.
Inbred Isonomy
3 and 4 rems.
Exp. Argentina 1915.
Moderate success.

(Dolabella produced:)

MYROBELLA
1930
Tetratema
won 11 races
and £16,100
bred 6 wnrs.

(Myrobella produced:)

SNOWBERRY
1937
Cameronian
won 2 good races,
bred 9 wnrs.

***† BIG GAME**
1939 2, Ch.
Bahram
Inbred White
Eagle 3 and 4 rems.
Sired Queenpot 1,
Ambiguity O,
*Combat and the
dams of Arctic
Explorer E
*Hethersett L,
and Sodium IRD,
L.

(Snowberry produced:)

***† CHAMOSSAIRE**
1942 L,
Precipitation
Sired Cambremer L,
Santa Claus D, IRD
& the dam of Khalkis
E.

ARIANA 1951
Tehran
did not win

SNOW BLOSSOM
1957
Flush Royal
did not win

SNOW KNIGHT
1971 D,
Firestreak
exp. Can., won 6
out of 9 starts in
USA in 1975, to
stud in Canada
1977.

MINORU

1913 SERENISSIMA
won 2 small races,
bred 8 wnrs. of 40
races – £47,800

SELENE
1919
Chaucer
won 15½
races val.
£14,386.
Bred 10
wnrs. of 30
races and
£47,300.

***†HYPERION**
1930 D, L,
Gainsborough
Inbred St. Simon
3 & 4 rems.
& Galopin at
4th. The superb
little (15.1½hh)
horse through
which alone the
line of Hampton
has survived and,
thanks to his son
Aureole, has spread
to the extent that it
might well endure
for some time to
come. His close in-
breeding to St.
Simon is easily
explained without
a chart: –
His sire Gainsborough was out of the Oaks winner, Rosedrop by St. Frusquin (by St. Simon) and his dam, Selene, was by St. Simon's son Chaucer. Hyperion won 3 times as a 2yo and started his 3yo career by winning the Chester Vase; in addition to the Derby and Leger he won once more at 3 and twice again as a 4yo. He was beaten a long way into 3rd place in the Gold Cup. A very good racehorse but not a super one, he was in fact beaten 3 times, but never out of a place. He sired the winners of 17 classics and his daughters bred the winners of 9; his influence was great in USA through his sons ALIBHAI, Heliopolis, Khaled and Seven Seas, and in Australia through High Peak (p. 16).

ALL MOONSHINE
vis. Fr. '51,2, USA
'56, '60.
1941 Bobsleigh
won once, bred
5 wnrs.

***†MOSSBOROUGH**
1947 Nearco
Inbred Chaucer 3 & 4
rems. Won 5 mid.dist.
races and £4,606. Sired
*†Ballymoss L, AT, E,
GVI, IRD, Noblesse O
and the dams of
*Exbury AT and
Sagaro GP, GC (3 times)

TRANQUIL & her bro. **†SCHIAVONI**
1920 1, L,
Swynford

†SCHIAVONI
1922 Swynford
a mod. mid. dist.
h'cpr., but sire
of Schiaperelli
(see later)

COMPOSURE
1924
Buchan
won 3 small
races, bred
4 wnrs.

COMPLACENT
1930
Colorado
won twice,
bred 5 wnrs.

LACKADAISY
1940
Felstead
won once,
bred 4 wnrs.

LACKADAY
1947
Bobsleigh
won 3 races
at 2y, bred
5 wnrs.

***KALYDON**
1956
Alycidon
Inbred Gains-
borough 3 &
4 rems. Sired
Park Top GVI

***†FAIR COPY**
1934
Fairway
Inbred to Cyllene
at 4th rem. Won 8
mid.dist.races and
£10,500. At stud
in France 1939-48,
then in UK, exp.
Sweden 1956.
Sired Orfeo GP,
*Sayani,
†Menetrier and
the dam of
*Tanerko.

***BOSWORTH**
1926 GC,
Son In Law
2nd in Leger
by a sht.hd.
Sired Boswell
L,E, Plassy &
the dam of
Gunshot (USA)

SWYNFORD

1915 FERRY 1 & her bro.***†SANSOVINO**
failure at stud,
exp. Poland
1930

***†SANSOVINO**
1921 D,
The mare, Pilgrim-
age (p.37, 167, 168)
appears twice in
their 3rd gen. San-
sovino won 6 races
in all and £17,700.
A disappointment at
stud considering that
he had the run of
all the best mares.
Sired Sandwich L,
the obscure Jacopo
and the dams of
*†Tudor Minstrel
2, Zarathustra GC
and Eble.

STEADFAST

1923 PIAZETTA
won 3 races,
bred 3 wnrs.
Exp. France 1935

HARBOUR LIGHT
1937
Pharos
won once

HERODIADE
1943
Tourbillon
won 1 good race

THUNDERHEAD II
1949 2,
Merry Boy
exp. S. Africa.
His sire MERRY BOY
1938 won 4 races up
to 12f in France, he
was by Asterus from
a Cyllene mare and
inbred to that horse
at the 3rd and 4th rem.
Imp. UK 1952.

ST. ASTRA 1904 FO

The origins of **ST. ASTRA** 1904 FO. By *†**LADAS** 1891 2, D, and 2nd in the Leger, a good racehorse, beautifully bred, by Hampton out of Illuminata (p.39), but disappointing as a sire. Besides St. Astra, he got 3 not very important classic winners and the dams of 4 reasonable stallions and a 1000 gns. winner.

Ex ST. CELESTRA 1897	by St. Angelo (p. 77), won 2 small races, exp. from UK to France 1903.
Ex ASTROLOGY	by Hermit.
Ex STELLA	by Brother To Strafford.
Ex a bay mare 1861	by Toxophilite. Ancestress of *Bull Lea (USA) and of Gregalach, the 1929 Grand National Winner.
Ex MAID OF MASHAM	(p.11). Family no. 9.

St. Astra is inbred to Newminster at the 4th remove.

HER MATINGS:

LE SAGITTAIRE 1892, a useful racehorse but said to lack quality. By the French champion sire, Le Sancy (grey) from a grey mare by Doncaster, he was a chestnut; his granddam, another grey, was by the grey Strathconan; so was Le Sancy's grey dam. Thus Diavolezza, herself a chestnut, had 2 strains of the not very notable Strathconan at her 4th remove. **VERDUN** 1906 GP, (F2), a good racer, but Astrella (next page) seems to be his only claim to fame while at stud in France; he went to Argentina in 1918. He was by Rabelais (by St. Simon) from a mare by the unfashionable, despite getting a FO winner, Gournay. Astrella was inbred to Galopin at the 4th remove. *†**SARDANAPALE** 1911 FD, GP, a very great racehorse indeed. He was straight in front, a fault which he was inclined to transmit to his offspring; they were sometimes very difficult to train. By Prestige, who had passed on *his* straightness, out of a beautifully bred mare (sis. to Vedas 2) by Florizel II (by St. Simon), his granddam was by Rosicrucian. He sired: Doniazade FO, Fiterari GP, (F2, FL), †Apelle (It.D) and, greatest of all, the mare Zariba (see later). The Rosicrucian in him gave Titanite two of that strain in her 4th generation.

ST. ASTRA 1904 FO.

(whose dam had come from UK in 1903)

Mated with: –

Produce: –

LE SAGITTAIRE

1911 DIAVOLEZZA (F1)

- FARIZADE 1921
  Sardanapale
  - KASSALA 1926
    Cylgad
    won twice
    - CHEVALERIE 1933
      Abbot's Speed
      won twice flat
      and do Hurdles
      - ***†PRINCE CHEVALIER 1943 FD, (Lupin)**
        Prince Rose
        Won also 5 other big French races and was beaten a sht. hd in both GP and AT. Foaled in France, imp. UK 1947, his first crop produced the Derby winner Arctic Prince, to be followed by Beau Prince II, Pirate King, *Court Harwell, *Doutelle, *Charlottesville FD, GP, Soltikoff AT and La Paiva, dam of Brigadier Gerard.
  - AZIYADE 1938
    Herbalist
    - ANNA COMNENE 1942 Fair Copy
      - COMMEMORATION 1953 Vandale
        - **ARMISTICE 1959 GP,**
          Worden
- CELERINA 1928
  exp. Fr. '38
  Teddy
  - **PENSBURY 1940 GP, (Lupin)**
    Brantome

VERDUN

1912 ASTRELLA

- ***†ASTERUS 1923 Ch. (F2)**
  Teddy
  Also won the Hunt Cup carrying top weight. Became a great sire of females, the two most important being **Adargatis FO and Astronomie** (p.12, App. D). Other daughters of his bred Djeddah E,Ch., Scratch II FD,L, and Talma II L. Also the following stallions were his sons: – *Abjer, *†Astrophel, *†Atys and *Merry Boy – but none of them were able to maintain for him a place in the Bend Or dynastic succession. Asterus was inbred to Hampton at the 4th remove.

SARDANAPALE

1917 TITANITE
imp. UK '21
exp. France 1923

- TATOULE 1925
  Alcantara II
  - ***FASTNET 1933**
    Pharos
    A high class race-horse in India, from whence he returned to France and sired Le **Paillon AT**, Flocon E, and *Fast Fox.

QUEENIE 1904

The origins of **QUEENIE** 1904. By ***WAR DANCE** 1887 by Galliard. Exp. France as a yearling, there he won 12 races. His most far-reaching get as a sire was Roxelane FO, dam of *†Roi Herode (sire of The Tetrarch). He sired also the big winner and important *Perth FD, GP, (F2, FL) and Mordant FD. War Dance would have been known as a classic sire at the time of Queenie's conception, he was inbred to The Baron at the 4th remove. Queenie was foaled in France.

Ex QUILDA 1894	by **Gamin** 1883 (FL), by Hermit. Quilda was foaled in France and won there well; imp. UK 1908.
Ex QUICK THOUGHT 1884	by Forerunner, exp. France 1890.
Ex MAGNOLIA	by Lecturer.
Ex MAHONIA	by King Tom, a granddaughter of the famous Queen Mary 1843 (did not win) by Gladiator who had 17 foals of which no less than 6 daughters founded classic branches of family no. 10.

Queenie had only 6 foals.

HER MATINGS:

SEASICK 1905=FD, a tough horse who won a lot of good races in France; by Elf II GC from a mare by Le Sancy. He sired also the stallions Antivari and Battersea.

***PRESTIGE** 1903, never to be forgotten as the sire of Sardanapale (v.s. St. Astra), he was unbeaten in a lot of moderate races. He cannot have been beautiful for he was said to be straight in front and to have had bad hocks. His sire was the depressing Le Pompom who won one race and suffered from stones, his dam was by Reverend. Quenouille (next page) is inbred to The Scottish Chief at the 4th remove.

QUEENIE 1904 *another French Mare*

Mated with: – SEASICK

Produce: – 1915 QUARANTINE (dam also of Battleship, winner of the Grand National 1938)

QUOI 1920 FO, Chut

AUDE 1927 Clarissimus won 4 races

EN FRAUDE 1934 FO, Pharos

Mated with: – PRESTIGE

Produce: – **1916 QUENOUILLE FO**

FICELLE 1923 La Farina

VIELLE CANAILLE 1930 Zionist

VIELLE MAISON 1936 Finglas won 4 good races

***VIEUX MANOIR 1947 GP, Brantome** see p. 169

VIELLE PIERRE 1951 Blue Peter

HERMIERES 1958 FO, Sicambre

CRYSTAL PALACE 1974 FD Caro

and her sister

PAILLE 1927 La Farina

BALALA 1934 Fiterari

BALATAME 1943 Sirtam

BOBAR II 1958 Ch. Bozzetto
His sire *†BOZZETTO 1936, won 5 races in Italy. He was by Pharos from Bunworry by Great Sport and inbred to St. Simon at 4th rem. He sired also Tamanar FD.

and her sister

QUEEN'S ROBE La Farina

MADAME FIERE 1936 Blenheim

MADAME PATROL 1941 Prince Rose

FALLAHA 1950 Sayani

CECROPIA Alizier

BELLA MOURNE 1962 Mourne

RESCOUSSE 1969 FO, Emerson
Her sire EMERSON (imp. Fr. '63) was foaled in Brazil in July 1958 and was unbeaten there in 5 races. He was by Coaraze from a mare by Full Sail.

FRIZETTE 1905

The origins of **FRIZETTE** 1905. By ***HAMBURG** 1895 (USA), a stayer, champion sire USA. In the background of Northern Dancer's pedigree his daughter, Lady Hamburg, is the dam of Chicle and in Nijinsky's she is the dam of Dis Donc (sire of Bull Page's 2nd dam, see p. 155).

Ex ONDULEE 1898	by St. Simon. Ondulee was exp. USA 1903.
Ex ORNIS 1890	by Bend Or. Sis. to **Orion** Ch. (exp. Aust-Hung. 1906).
Ex SHOTOVER 1879, 2, D	by Hermit.
Ex STRAY SHOT	by Toxophilite. Founded one other family, as did her dam VAGA 1858 by Stockwell. Family no. 13.

One of the strongest bottom lines recorded so far.

Frizette was foaled in USA and exported to France in 1908.

She had more than 13 foals, these are her matings:

IRISH LAD 1900, bred in USA and won 12 good races there, then sent to France where he got a French Oaks winner and Banshee – not much else. His sire was an English horse, Candlemas, exp. to USA, a brother to St. Blaise D (by Hermit); his dam was American and not to be found in the English Stud Book.

***PRINCE PALATINE** 1908 L, GC (twice), E, also won the Doncaster and Coronation Cups and the Jockey Club Stakes. He stood in UK, France and USA and sadly he died through an accident aged only 16 (see p.89, 157, 166, App. C, App. D). He had not shown himself to be a good stallion but he did leave in France the stayer *Rose Prince who maintained his position in the St. Simon partriarchy. His sire was Persimmon (by St. Simon) and his dam an Isinglass mare. Princess Palatine (next page) was inbred to St. Simon at the 3rd remove.

SWEEPER II 1909 2, one of the earlier invaders from USA to capture an English classic. By Broomstick (one of the great Bonnie Scotland line in the States) from a mare (exp. from UK 1903) by Sir Hugo which, carrying Sweeper, was refused entry into France due to an administrative error and was disembarked in England.

†DURBAR II 1911 D, was also the sire of two of Frizette's granddaughters and the maternal grandsire of Tourbillon. Foaled in France, he was by Rabelais (by St. Simon) from an American-bred mare whose dam, Urania (ran in 87 races, won 35 and was placed 34 times) was by Hanover, Frizette's grandsire. Not being in the G.S.B. he presumably had little attraction for UK breeders. His ancestry makes Frizelle inbred to St. Simon at the 3rd remove and to Hanover at the 3rd and 4th. He was exp. USA 1926.

FRIZETTE 1905

(USA bred and came to France 1908)

Mated with: – IRISH LAD

Produce: – 1910 BANSHEE (F1)

- DURBAN 1918
  Durbar II
  (top 2yo France)
  - ***BANSTAR 1923**
    Sunstar
    won many races in France, sired d. of Le Volcan.
  - DIADEME 1927
    Ksar
    won well
    - ALBARELLE
      1932 Kantar
      - **CARAVELLE**
        1940 FO, (F1)
        Abjer
  - and her bro.
    (inbred to Omnium II 3 & 4 rems.)
  - ***†TOURBILLON**
    1928 FD, Ksar
    Three times champion sire in France and it is possible that he might have had more UK mares visit him had not his maternal grandsire, †Durbar II D, been excluded from the General (UK) Stud Book, whereas in France and USA he qualified. Thus, until this was rectified later on, Tourbillon was technically taboo in UK. The most important of Tourbillon's sons was *†Djebel 2, AT, (F2) (see p.180, App. D) and besides him he got the winners of 6 other classics and 6 top stallions. His daughters bred the winners of 5 classics and the good sire *†Nosca.
- and her sister
- HELDIFANN 1921
  Durbar II
  - DJEZIMA 1933
    Asterus
    - TOURZIMA
      1939
      Tourbillon
      - GLORIANA (sis)
        Pharis II
        - GAIA 1962
          Shantung
          - LICATA
            1969
            Abdos*
            - **ACAMAS**
              1975
              FD
              Mill Reef
      - COREJADA
        1947 (F1, IrO)
        Pharis II
        - **APOLLONIA**
          1953 FO, (F1)
          Djebel
          Inbred Tourbillon 2 & 3 rem.
        - **MACIP 1952**
          GC, (FL)
          Marsyas II
          Inbred Asterus 3 & 4 rem.
    - **PRIAM II**
      1941 Pharis
      won Hardwicke Stakes exp. USA, Sired Stratonice FO.
    - **DJEDDAH**
      1945 E, Ch.
      Djebel
      exp. USA '50 Inbred Teddy & Durbar 3 & 4 rems. Sired dam of *Never Bend see p. 152

**Abdos 1959 by Arbar ex Pretty Lady (d. of Dynamiter Ch (twice) by Umidwar). Ran only at 2y, unbeaten in 2 big 8f races inbred Teddy 3 and 4 rem.*

PRINCE PALATINE

1919 PRINCESS PALATINE
did not run
exp. USA

VALKYR 1923
Man o'War
bred 7 wnrs.

VALSE 1932
Sir Galahad III

VIENNA 1941
Menow

TWO CITIES
1948 Johnstown

***BALDRIC II**
1961 2, Ch.
Round Table
Inbred Sir Galahad III
4 rem. Sired Irish Ball IRD.

and her sis.

VAGRANCY
1939
Sir Galahad III
won 15 races

BLACK TARQUIN
1945 L,
Rhodes Scholar
Foaled USA, imp. UK, then to stud USA 1950 and to Ireland 1954
A notable sire of jumpers.

SWEEPER II

1920 ONDULATION

GILDED WAVE
Gallant Fox

ADORADA 1947
Hierocles

CHARMING ALIBI 1963
Honey's Alibi

DAHLIA 1970
GVI (twice)
(Wash. Int. Ir.O)
Vaguely Noble
Inbred Hyperion
4 rem.

DURBAR II

1922 FRIZELLE

ORLANDA 1929
Craig An Eran
won twice

CILLAS 1925
FD, Tourbillon
Inbred Durbar
3 rem. Exp. from France to Ireland 1939.
Sub-classic sire.

and his sister

ARRIBA 1942
Tourbillon

AURIBAN
1949 FD, Pharis
Inbred Durbar
4 rem. Won many other good races; said to have been difficult.

HONORA 1907

The origins of **HONORA** 1907. By GALLINULE p. 13, by now established as a classic sire.

Ex WORD OF HONOUR 1892	by Saraband p. 31. (THE SAME FORMULA THAT PRODUCED PRETTY POLLY – *BY* GALLINULE *EX* A MARE BY SARABAND). Word of Honour did not win, had 12 foals, none of note except this Honora, the only one by Gallinule.
Ex GEHEIMNIS 1879, O	by Rosicrucian. Geheimnis was exp. Germany '93, ancestress of Ramus FD, Jock II and Scratch II FD, L.
Ex NAMELESS	by Blinkhoolie.
Ex NO NAME	by Teddington and 7 generations back to Boadicea (p. 31) Family No. 14.

Honora had 15 foals to live, of these 6 won 12 races and £13,300.

HER MATINGS: Bayardo and Lemberg see Isoletta p.29.

*†**THE TETRARCH** 1911, darkish grey with astonishing white flecks and splodges, nick-named "The Rocking Horse". He raced only as a 2yo, winning all 7 races he ran in. His victory in the Coventry Stakes at Ascot is recorded as by 10 lengths and those writers who witnessed it consider that the Judge was far from generous. He suffered an injury in training early on as a 3yo and could not run again. Bred by the moderate French stayer, Roi Herode from a Bona Vista mare whose next 3 ancestresses were far from famous, his 4th generation contained two of each of the following notabilities: Doncaster, Speculum and Rouge Rose (dam of Bend Or and 3rd dam of Roi Herode). His career at stud lasted but a few years: by 1926 he was completely sterile. Of the 130 foals he got, 80 were winners. One of the last was the fantastic "Flying Filly" Mumtaz Mahal (see later) and, on account of her and perhaps the 2000 gns. winner, Tetratema, one cannot help associating his offspring with speed and possibly forgetting that he sired also 3 Leger winners.

*†**TRACERY** 1909 L, E, Ch. was foaled in USA and imported to UK as a yearling. He was by Rocksand from the Orme mare **Topiary** (exp. USA '06). In 1920 he was sold to Argentina but not before he had sired the Derby winner, Papyrus, 2 other classic winners, 3 very good stallions, the dam of Scuttle 1, the great mare Teresina (dam of Alibhai) and now, Clear Evidence (next page). Tracery returned to UK in 1923 but died the next year. He was inbred to Galopin at the 4th remove.

HONORA 1907

Mated with:–

Produce:–

- **BAYARDO**
  - 1914 BAYORA
    did not win
    bred 6 wnrs.
    exp. Fr. 1932
    - MISS HONOR
      1932
      Mr. Jinks
      - MARIE ELIZABETH
        1948
        Mazarin
        - CAT O'MOUNTAINE
          1967
          Ragusa
          - TAP ON WOOD
            1976 2
            Sallust
    - BACHELORA
      1927
      Bachelor's Double
      won 1 small race
      bred 4 wnrs.
      - SILVER BIRCH
        1933
        Blandford
        won once
        bred 8 wnrs.
        - HASTRA (exp. USA '52)
          1942 Hyperion
          won once
          bred 4 wnrs.
          - ROSE OF YERODA
            1948 Nearco exp. France '52 won twice at 2y. exp. USA 1956
            - **ROSE ROYALE II**
              1954 1, Ch.
              Prince Bio

- **THE TETRARCH**
  - 1916 HERODIAS
    exp. USA, won 5 races there.
    - HERODIADE
      Over There II
      - BANISH FEAR
        1932
        Blue Larkspur
        - NOT AFRAID
          1948
          Count Fleet
          - **PRINCE JOHN**
            1953
            Princequillo
            Sired dam of Caracolero 1971 FD.
    - ***PILATE (USA)**
      1928
      Friar Rock
      Sired Eight-Thirty.

- **LEMBERG**
  - **1918 LEMONORA**
    GP, ran 3rd in Humorist's Derby. Closely inbred to Isonomy, having one strain at the 3rd and two at the 4th remove. He was no good at stud.

- **TRACERY**
  - 1920 CLEAR EVIDENCE
    won 1 small race
    bred 4 winners
    - **ARTIST'S PROOF**
      1926 Gainsborough
      Won 8 good mid-dist. races. Exp. France 1933 and Hung. '38. Sired Fine Art and the dam of *† Petition E, see p. 158

SAPERLIPOPETTE 1909

The origins of **SAPERLIPOPETTE** 1909. By **CHERI** 1898 GP, (F2), a grandson of St. Simon by **St. Damien**, from a mare by Wellingtonia. Sired also Juveigneur. Cheri was said to have been extremely handsome, perhaps inherited from St. Damien who was supposed to have been of perfect conformation.

Ex SATANELLA 1902	by **ERMAK** 1888 FD, also 2nd in GC. Satanella seems to be his only claim to fame.
Ex STORMY PETREL	by Gallinule. Exp. France 1898, sis. to Wildfowler L.
Ex TRAGEDY	by Ben Battle.
Ex THE WHITE WITCH	by Massinissa.
Ex JEU DES MOTS	by King Tom, one of the 5 family-founding daughters of **JEU D'ESPRIT** 1852 family no. 7.

Saperlipopette won 4 races.

She bred 10 winners.

SAPERLIPOPETTE'S MATINGS:

Rabelais, being a son of St. Simon, endowed Shocking with 2 strains of that scion, in her 2nd and 4th generation.

*†**RIALTO** 1923, a middle-distance performer who ran 37 times and won 17 races, was by Rabelais out of a well-connected mare by Helicon (by Cyllene). Rialto sired *†Wild Risk, †Herne The Hunter, Rienzo and (next page) the dam of Sicambre.

SAPERLIPOPETTE 1909

(whose granddam came from UK to France)

Mated with: – SWYNFORD | RABELAIS | RIALTO

Produce: –

SWYNFORD

SHAMELESS

LADY CHATTERLEY
1932 Rialto
won twice

LAIS 1940
Fantastic

LANDERINETTE
1952 Sicambre

LA LAGUNE
1965 O,
Val De Loir
Inbred to Rialto
4 rem.

RABELAIS

1919 SHOCKING

SUAVITA 1928
Alcantara II

SIF 1936 Rialto
won 3 good races

***†SICAMBRE**
1948 FD, GP,
Prince Bio
Inbred Rabelais
3 & 4 rems.
won also 6 other
good races. Sired
Sicarelle O,
Hermieres FO,
Belle Sicambre
FO, *Tiziano
(It.L), Phaeton GP,
*†Shantung,
*Celtic Ash and
*Diatome.
(p. 102)

and his sis.

SENONES 1952
Prince Bio

***SNOB 1959**
Mourne
exp. Japan 1972
Sired Goodly FD,
His sire, *†MOURNE
1954, a miler, by
Vieux Manoir from
a Nasrullah mare,
sired also Rock Roi
and the dam of
Rescousse FO.
Mourne was imp.
UK '64 and ret.
France '72.

RIALTO

1932 M'SELLE SALLE
won 4 mod. races

M'SELLE PETIT PAS
1939
Blue Skies
won 1 good race

M'SELLE PAGANINI
1950 Loliondo

PASQUINADE II
1957 Vandale

PHAETON 1964
GP, Sicambre
Sired Vitiges 1973 Ch.

BILL AND COO 1909

The origins of **BILL AND COO** 1909. By WILLIAM THE THIRD (p. 73), who, at the time of this mating, had yet to make his name as a sire.

Ex COOEE 1899	by †**TRENTON** 1881, a stayer, won 8 out of 14 races in N.Z. Imp. to UK in 1896 aged 15. Four years later he sired Rosaline the dam of Rosedrop O, (immortal as Gainsborough's dam). Trenton also sired †Torpoint (won the Alexandra Stakes twice) who got the dam of Buchan and Saltash (p. 66). Cooee won 2 mod. races, exp. France 1913 bred 4 winners, inc. **Call Of The Wild** 1913 (exp. Argentina '24) sire of dam of April The Fifth D.
Ex LADY STERLING	by Silver.
Ex ZORILA	by Nuneham
Ex GUADELOUPE	by Neptunus, granddaughter of Tasmania 1854 whose daughters founded 2 classic lines. Family no. 2.

Bill And Coo won 2 good races.

She had 12 foals to live, 5 of which won 22 races of no great note.

BILL AND COO'S MATINGS:

Mated with †**CORCYRA** 1911, won 8 middle distance races and £10,100, was beaten a sht. hd. in the 2000 gns. By Polymelus out of the good mare **Pearmain** by Persimmon, he was inbred to Hampton at the 3rd and 4th removes and to Galopin at the 4th. Corcyra unfortunately died aged 9 but not before he had sired the dams of Commanderie FO, GP, *Deiri (FL), Ballyferis and Pompey as well as the important Cinq a Sept (next page).

BILL AND COO 1909

Mated with: – THE TETRARCH

Produce: – 1917 TETE A TETE
exp. France '37
won 4 good races
bred 4 winners.

- BOOKTALK
  1926
  Buchan
  won once
  bred 4 wnrs.
  - BORROW 1931
    Tremola
    won 5 races
    - LOAN 1940
      Portlaw
      did not win
      - MARTIAL LOAN
        1950
        Court Martial
        did not win
        bred 4 wnrs.
        - SANS LE SOU
          1957 Vimy
          exp. France 1964
          won once, bred
          5 wnrs.
          - ***BUSTED 1963**
            E, GVI,
            Crepello
            Sired Weaver's
            Hall IRD and
            Bustino L.
  - SERIAL 1932
    Solario
    won once at 2y
    - ***BALLYOGAN**
      1939 Fair Trial
      Inbred Sundridge
      4 rem.
      An Irish sprinter,
      sired Ki Ming 2
      and the dam of
      *Le Levanstell
- CONVERSATION
  PIECE 1935
  Orpen
  won Ir. Oaks
  - SOCIETY'S VOTE
    1944 Wyndham
    - SOCIETY'S WAY
      1959 Kingsway
      exp. Peru 1955
      - PAMPELONA II
        1966 Postin who
        won Peruvian Triple
        Crown
        - **EMPERY 1973**
          D, Vaguely Noble

Mated with: – CORCYRA

Produce: – 1919 RACKETY COO
did not win,
bred 4 wnrs – £11,900

- CINQ A SEPT
  1924 Roi Herode
  won 6½ good races
  exp. France 1929
  - ASSIGNATOR
    1930 Teddy
    - CINQUEPACE
      1934
      Brown Bud
      - IMPERATRICE
        1938 Caruso
        - SOMETHING
          ROYAL
          1952
          Princequillo
          - ***SIR GAYLORD**
            1959 Turn To
            Inbred Pharos
            4 rem.
            Sired SIR IVOR
            2, D, Ch. see
            p. 60
          - **SECRETARIAT (USA)**
            1970 (Preak. Bel.)
            (Ken. D.)
            Bold Ruler
  - TWILIGHT TRYST
    1935
    - BALLA TRYST
      1940 Balladier
      - BANK ACCOUNT
        1948 Shut Out
        - JAMBO 1959
          Crafty Admiral
          - SEXIMEE 1966
            Hasty Road
            - **NONOALCO 1971**
              2, Nearctic
  - **GENTLE TRYST**
    Sir Gal. III
    - UP THE HILL
      1938 Jacopo
      - **TUMBLING 1953**
        War Admiral
        - DETERMINED
          LADY 1959
          Determine
          - PRINCESS POUT
            1966 Prince John
            - **ALLEGED 1974**
              AT twice, 3 & 4y
              Hoist the Flag

CATNIP 1910

The origins of **CATNIP** 1910. By *†SPEARMINT 1903 D, GP (p. 73), not yet renowned as a sire.

Ex SIBOLA 1896, 1,	by **THE SAILOR PRINCE** 1880, a not very good handicapper who won the Cambridgeshire. Imp. from USA 1897 also ran 2nd in Oaks by a head.
Ex SALUDA 1883	by Mortemer.
Ex PERFECTION	by Leamington.
Ex MAIDEN	by Lexington, an American branch of family no. 4 which went there with a Cub mare in 1762.

Catnip was bred in Ireland, won one small race and went to stud in Italy 1918.

HER PEDIGREE IS ENTIRELY FREE OF ST. SIMON.

She had 6 other foals to live, none of the class of the two overleaf.

CATNIP'S MATINGS:

Tracery (p. 93) gives Nera Di Bicci St. Simon and his sister Angelica in her 4th generation.

***†HAVRESAC II** 1915, bred in France. Won 9 races; became champion sire in Italy. He was by Rabelais, so his grandsire was St. Simon, and, since his granddam was also by St. Simon, he was inbred at the 2nd and 3rd removes. Thus the lack of St. Simon referred to above in Catnip's pedigree was amply made up in her daughter Nogara's.

NOGARA. From the 3 tremendous sons of Nogara recorded here, one can imagine the countless times her name appears in the pedigrees of winners all over the world; nevertheless, these notes do not begin to do justice to her enormous influence. Therefore, outside the scope of this book, here is a brief sketch of the performances of her other offspring: **NICOLAUS** 1939, by Solario, won 9 Italian sub-classic races, went to stud in Ireland and got, among quite a lot of unimportant winners on the flat, the winner of the 1961 Grand National, Nicolaus Silver. **NICCOLA D'AREZZO** 1943, by Ortello, won moderate races in Italy and became a poor sire in USA. **NAUCIDE** 1945 by Bellini, (closely inbred to Havresac II), won 6 good races in Italy, went to stud in UK in 1949, then to France in 1956. No classic connections so far. KING OF TARA 1947, by Torbido, won nothing and was sent to S. Africa. She had only one daughter, the promising **NERVESA** 1941, by Ortello, who won the Italian Oaks and another good race. She bred a useful horse, **Nomellini** 1946, by Bozzetto, and he, at stud in UK from 1950 until he died in 1958, has done nothing so far to make Nervesa a classic ancestress; however, in June '77, *Fortino II (ran only at 2y and sent to Japan) has entered into classic reckoning by siring *Caro (p. 36). Fortino was a great grandson of Nervesa.

CATNIP 1910

(An Irish mare who went to stud in Italy 1918)

Mated with: – TRACERY

Produce: – 1918 NERA DI BICCI won 8 races

NUVOLONA 1926 Hurry On

NAVARRO 1931 Michelangelo Inbred Spearmint 3 & 4 rems. Sired dam of Toulouse Lautrec. His sire MICHELANGELO 1918, Italian, was by Signorino from the Italian D & O winner Fausta by Spearmint. Michelangelo also sired *Zuccarello.

NANNOCIA 1927 (It. 1) Michelangelo

NANON 1935 Graf Isolani did not win bred 8 wnrs.

NIXE 1941 Arjaman won 3 races bred 9 wnrs.

NECKAR 1948 (Ger. D) Ticino Inbred Dark Ronald 3 & 4 rems. Sired the dam of Pia O. His sire *†Ticino 1939, German D winner and champion sire, by the German D winner Athanasius from a mare by Aditi, sired also Bella Paola 1,O,Ch.

Mated with: – HAVRESAC II

Produce: – 1928 NOGARA won Italian 2 & 1 & 11 other good races, bred 8 big winners from 8 foals. Imp. to Ireland 1946

***† NEARCO fld. Italy. 1935 GP, Pharos. imp. '38** Inbred St.. Simon 4 rem. He won also 13 big races in Italy and was never beaten. He sired the winners of 9 classics and his daughters were dams of the winners of 12. He ranks now as the founder of his own dynasty which bids fair to survive beyond all others; this he has achieved through his sons Nasrullah and Royal Charger in USA, Nearctic in Canada and Mossborough (hopes pinned on Ballymoss and Royal Palace) in UK. see p. 83

***† NICCOLO DELL 'ARCA 1938 (It.2,D) Coronach** won also 9 other big Italian races; stood there from 1942 to '48 when he came to UK. None of his sons have reached classic status, but his daughters were dams of Psidium D & of the wnrs. of 3 long-distance classics. His sire *†CORONACH 1923 D,L,E, won 10 out of 14 races, the Derby by 6 lengths and the Leger in record time. He was unsound in his wind. By Hurry On out of Wet Kiss by Tredennis, as a sire his great successes were limited to abroad; in addition to the very influential Niccolo (above) he got Corrida AT (twice) (see later), Cranach 1938 (a p.o.w. of the Germans who later became a good stallion in France) and his daughters accounted for *†Coaraze FD and *†Alizier. Coronach was exp. NZ 1940

†NAKAMURO 1940 Cameronian won 4 Italian races; stood there from 1944 to '49 when he went to Ireland and then back to Italy. '54. Sired the dams of Molvedo AT and Yrrah Jnr. (sire of Hurry Harriet Ch.) His sire *†CAMERONIAN 1928 2, D, Ch. was by Pharos out of a Gainsborough mare and he had two strains of St. Simon in both his 4th and 5th remove. He went to stud first in England then to Argentina in 1941 where he died 14 years later. In addition to Nakamuro he sired Scottish Union L and Finis GC neither of which did him or his ancestry justice in their stud careers. His daughters produced the stout *†Chamossaire L and Fighting Charlie GC.

REINE MAB 1912

The origins of **REINE MAB** 1912. By RABELAIS see p. 24.

Ex JUSTICIA 1896	by *†**LE SANCY** 1884, ran in France for 5 seasons winning 27 out of 43 races over middle distances. He was 4 times champion sire of France and his fee corresponded to St. Simon's, 500 guineas. His offspring won 5 classics, and his daughters bred 6 classic winners. His sire was the 2000 winner **Atlantic** and his dam was by Strathconan; both his parents were exported from UK to France. He was inbred to Windhound at the 3rd and 4th rems. Justicia was a great 2yo; dam of **St. Just** 1907 (by St. Frusquin) who sired the dams of Mont Bernina, *Telerifique, Godiche and the 2nd dam of *†Mieuxce FD, GP.
Ex THE FRISKY MATRON 1879	by Cremorne. Exp. from UK to France 1882.
Ex MAY FAIR	by The Marquis.
Ex THE MAY QUEEN	by Trumpeter, family no. 12.

Reine Mab bred 11 winners.

HER MATINGS:

†**SANS SOUCI II** 1904 GP, (Lupin), by Le Roi Soleil GP from **Sanctimony** by St. Serf. Although he sired *†La Farina (below), Sans Le Sou and Cadum, his daughters made him more famous by breeding 5 classic winners, †Mousson and, above all, *VATOUT (p. 166). Lanette (next page) was inbred to St. Simon at the 3rd and 4th removes.

*†**LA FARINA** 1911 (Lupin) and 2nd by a neck to Sardanapale in the GP. By Sans Souci II (above) from an Isinglass mare, his granddam being by St. Simon he was inbred to that scion at the 3rd and 4th removes, which meant that the sisters Broceliande and Ygerne (next page) had a St. Simon strain in their 3rd, 4th and 5th removes. La Farina sired 2 classic winners, *Bubbles and 3 other useful stallions; his daughters bred, as well as Maurepas, †Verso II FD, AT, (FL), *†Astrophel, *†Vandale and *†Meridien

CADUM 1921, a staying rogue, won the Cadran and other big races, was disqualified in favour of Priori in the AT. By Sans Souci II from the dam of Bubbles by Neil Gow, he sired the 2nd dam of Prince Bio.

REINE MAB 1912

whose Granddam came from UK to France

Mated with: –

Produce: –

SANS SOUCI II
- 1923 LANETTE
  - NANAIA 1929 Kircubbin
    - NYMPHE DICTE 1935 Diolite
      - LA DAME BLANCHE 1943 Biribi
        - ALBA NOX 1951 Coaraze
          - **WHITE LABEL 1961 GP, Tanerko**
      - DICTAWAY 1952 Honeyway
        - ***DIATOME 1962 (Wash. Int) Sicambre** had the misfortune to be foaled the same year as Seabird II. So far Diatome has sired STEEL PULSE IRD (exp. Australia 1974)

LA FARINA
- 1925 BROCELIANDE
  - ***MAUREPAS 1937 GP, (Cadran) Aethelstan** Sired Camaree 1 and her sis. Marmelade, granddam of *Seabird II D, AT
- and her sister 1926 YGERNE
  - ANATH 1932 Massine won 3 races, bred 4 wnrs.
    - LADY MACBETH 1944 Telerifique won ½ race
      - BRIEF CANDLE 1950 Brantome won once
        - **TIMANDRA 1957 FO, Court Martial**

CADUM
- 1929 FEE ESTEREL
  - ASHERATT 1935 Sunny Trace
    - ASHLEEN Alizier
      - ASTUCE 1964 exp. USA '68 Vieux Manoir
        - **IVANJICA 1972 AT**

LADY JOSEPHINE 1912

The origins of **LADY JOSEPHINE** 1912. By SUNDRIDGE see p.11, App. C, App. D.

Ex AMERICUS GIRL 1905	by **AMERICUS** 1892, a sprinter imp. to UK from USA in 1895, he was by Emperor Of Norfolk (grandson of Lexington) of a line which goes back in tail-male to Herod via Diomed, the first English Derby winner: **DIOMED** 1777-99 (by Florizel by Herod) won the first Derby – 1 mile then – and later on survived the voyage to America. There he was champion sire 7 times. His direct male line, so far as this book is concerned, ends with Americus, 7 generations and 115 years. I'd like to think that Mumtaz Mahal and all the nobility that stems from Americus Girl and Lady Josephine could just contain a trace of that very enduring Derby winner of nearly 200 years ago. Americus Girl won 12 races and £8372, bred 6 wnrs. of 32½ races.
Ex PALOTTA	by Gallinule. Palotta bred 8 wnrs. of 40½ races.
Ex MAID OF KILCRENE	by Arbitrator.
Ex QUERIDA 1870	by King John, Great granddam of Lesterlin p.15.

Family no. 9.

Lady Josephine was a first foal, she won 4 good races including the Coventry Stakes.

She had 7 foals of which 4 won 12 races and £22,500.

LADY JOSEPHINE'S MATINGS:

***†SON IN LAW** 1911, a fine staying "cup" horse who won 8 races including two Jockey Club Cups, a Goodwood Cup and the Cesarewitch, he had no classic pretentions. Of the line of Hamptön, by Dark Ronald from Mother-In-Law by Matchmaker, his 4th dam was Seclusion, dam of Hermit. He was inbred to Blair Athol at his 4th remove. Such was the combination that led to so many of the stayers in Europe in the last 50 years. The first of his classic progeny was the Oaks winner Straitlace who bred the sires Lovelace and Interlace and went to France in 1935; then followed three winners of 4 Gold Cups – Foxlaw, Bosworth and Trimdon; then Rustom Pasha E, Ch. and, of nine other effective stallions, I shall name only Winalot, Apron and Epigram. His daughters bred the winners of 3 classics and 11 of their sons proved notable stallions. Having thus pronounced Son In Law the supreme transmitter of staying-power, his mating with Lady Josephine goes most of the way to give the lie to that utterance. Thanks, I believe, to the influence of Sundridge (v.s.p.11, App. C, App. D) and her maternal grandsire, Americus, her inherited

LADY JOSEPHINE 1912

Mated with: – SON IN LAW; THE TETRARCH

Produce: –

- **SON IN LAW** → 1919 LADY JUROR won 3 races and £8057. Bred 8 wnrs of 28 races – £26,700
  - JURISDICTION 1925 Abbot's Trace won 4 good races at 2y.
    - BISHOPSCOURT 1945 His Grace did not win, bred 7 wnrs.
      - QUEEN OF SPEED 1950 Blue Train won twice, bred 5 wnrs.
        - ***KASHMIR II 1963 2, exp. Fr. '66 Tudor Melody** inbred Hyperion 4th rem.
  - & her bro. **THE BLACK ABBOT 1926 Abbot's Trace** ran only at 2y, won Gimcrack, sired Wychwood Abbot Ch. (twice) His sire *† ABBOT'S TRACE 1917 won 6 good mid-dist. races, by Tracery from a mare by Rightaway, sired also **The Macnab & Abbot's Speed.**
  - **THE RECORDER 1927 Captain Cuttle** won 6 races, mostly 1m. exp. Russia 1936, sired Poker Chip, d. of Queen Pot 1 & 3rd d. of *Petingo. His sire CAPTAIN CUTTLE 1919 (exp. It. '28) D was by Hurry On from a mare by Cyllene, exp.Italy. Sired also Scuttle 1, Pilade (It.D) & the dam of Tiberius GC.
  - **RIOT 1929 Colorado** ran only at 2y, won once bred 4 wnrs.
    - **COMMOTION 1938 O, Mieuxce** bred 8 wnrs.
      - ***COMBAT 1944 Big Game** unbeaten in 9 races, sired Agressor GVI
  - ***†FAIR TRIAL 1932 Fairway** A miler, won 7 races; died aged 26, his progeny did not go more than 10f. He sired the winners of three 2000s, one 1000, a Ch. & 2 Eclipses. The most important of his sons were Court Martial and Petition, each of them once champion sire like their father. His daughters bred *†Supreme Court GVI, Premonition L, Meld 1, O, L and Taboun 2. see p. 67
  - EVEN SCALES 1931 Salmon Trout exp. France 1934
    - ***EBLE 1940 Davout** Successful sire in France, got 2nd dam of Dhaudevi GP (FL)
  - SANSONNET 1933 Sansovino a sprinter, won 3 good races, bred 10 wnrs. of 36 races val. £42,000.
    - NEOLIGHT 1943 Nearco won 7 races
      - MINARET 1948 Umidwar won 4 races, bred 3 big wnrs.
        - **OMMEYAD 1954 (Ir. L) Hyperion** maternal grandsire of Giacometti Ch.
    - ***†TUDOR MINSTREL 1944 2, Owen Tudor** Unbeaten at 2y over 5f winning the Coventry (then 5f) & the National Breeders' Produce Stakes, ending up top of the Free Handicap. At 3y he won once at 7f and three times at a mile, including the 2000 and the St.James' Palace Stakes, 9 races in all, val. £24,600. Exp. USA 1959. Sired *King of The Tudors E and *Tudor Melody (p. 41)
- **THE TETRARCH** → 1921 MUMTAZ MAHAL founded a family of her own, see later.

speed over-rode her mate's propensity for long-distance. Look at Lady Juror's mates, two Derby winners, Abbot's Trace (mid-dist. wnr. but raced up to 14f), Colorado 2, E, 3rd in the Derby and sire of Felicitation GC (Colorado could be the exception) and the Leger winner Fairway, and note – I repeat with the exception of Riot – the reluctance of the descendants to stay more than 10f. Had the writer appreciated this in 1947 he would not have sunk his War Gratuity on Tudor Minstrel in the Derby.

SUN WORSHIP 1912

The origins of **SUN WORSHIP** 1912. By SUNDRIDGE p. 11.

Ex DOCTRINE 1899	by AYRSHIRE p. 35. Doctrine won Coronation Stakes.
Ex AXIOM 1888	by †Peter (by Hermit).
Ex ELECTRIC LIGHT 1876	by STERLING. Ancestress, through Axiom's elder sister, Petrel, of the stallions Sir Archibald, Roseland and Tolgus.
Ex BEACHY HEAD	by Knight Of St. Patrick and back 4 generations to Augusta 1818 O. Family no. 26.

Sun Worship did not win. She was inbred to Hermit at the 4th remove.

She had only 7 foals and 4 of them won 21 races and £27,500.

SUN WORSHIP'S MATINGS:

VOLTA 1912 a sprinter whose progeny won over a mile and more. Exp. USA 1924. Sired Franklin 1918 Ch. and the 3rd dams of Homeward Bound O and Zarathustra GC.

***†GAINSBOROUGH** 1915 2, D, L, GC (3yo) – a war-time GC run at Newmarket, but that doesn't alter the fact that he was unquestionably the best of his year and one of the best in this century. By the nearly all-conquering Bayardo (p. 30) from the Oaks winner Rosedrop by St. Frusquin, he was inbred to Galopin at the 3rd and 4th removes. His sons won 7 classics between them and his daughters bred the winners of 6, including Cameronian and Mahmoud. Apart from Hyperion, his even more potent lineal successor, six of his sons became stallions of consequence. Solario and his sister Imagery are inbred to Hampton at the 4th remove, they also carry three strains of Galopin at the 4th and twice at the 5th.

SUN WORSHIP 1912

Mated with:– VOLTA

Produce:– 1920 VOLEUSE won 3 races, bred 9 wnrs. – £19,000 exp. France 1934

VEILLEUSE 1924 Alcantara II exp. France 1926
- RED FLAME 1928 Vermilion Pencil
  - **CASTEL FUSANO 1935 Ksar** Inbred Omnium II and Alcantara II 3 & 4 rems. Sired 2nd d. of Hula Dancer 1, Ch. and Spy Well (sire of Flossy Ch.)
  - FLYING COLOURS 1939 Massine
    - ***†FONTENAY 1946 Tornado** Good wnr. in France, sired Tropique E & the dam of Nelcius FD.

CHOR BAZAR 1931 Gainsborough won 1 small race, exp. France 1939
- CHIKOO 1935 Firdaussi foaled France
  - OOLA HILLS 1945 Denturius bred 5 wnrs.
    - NELLIE PARK 1957 Arctic Prince
      - **b.f. PARK TOP 1964 GVI Kalydon** won 9 races in 4 seasons and £63,900.

BALA HISSAR exp. Braz. '42 1933 Blandford Top 2yo, sire of dam of Souepi GC

BAKOU 1936 Blenheim
- BADINE 1943 Finglas won twice
  - BOUCLETTE 1948 Victrix did not win
    - **BALTO 1958 GP, GC, Wild Risk** Inbred Swynford 4 rem.

Mated with:– GAINSBOROUGH

1922 *†SOLARIO L, GC, won 6 races and £20,900, champion sire of winners once and of broodmares 3 times. Although he sired two Derby winners, Midday Sun and Straight Deal, also 5 other stallions of importance, none of them have kept him as a progenitor of the line of Hampton; this was left to Hyperion, also by Gainsborough. Solario's daughters bred the winners of 3 Legers and 2 Champion Stakes, but the greatest among the twelve horses of consequence they produced was Royal Charger, sent to USA, see p. 126

& his sister 1923 IMAGERY vis. Fr. '33 won ½ race, bred 3 wnrs. of 10 races – £12,500.
- CAPTAIN'S FANCY 1928 Capt. Cuttle did not win, exp. France 1933
  - JOLIE BRISE Bubbles
    - CAPRICCIO 1945 Crapom
      - **SERIA 1949 FO, Norseman** her sire *†NORSEMAN 1940 won 9 mid.dist. races in France; by Umidwar from Tara (granddam of Worden) by Teddy; he sired also Montaval GVI, Free Man & the dam of Dushka FO. Norseman was exp. to USA in 1954 and returned to France in 1956.
- ***PHIDEAS 1934 (Ir. 2, D) Pharos** Sired Frieze O.

TILLYWHIM 1912

The origins of **TILLYWHIM** 1912. By MINORU p. 81.

Ex LILY ROSE 1907	by *†**WILDFOWLER** 1895 L, by Gallinule ex **Tragedy** by Ben Battle (exp. France 1908). Lily Rose won Gimcrack and 2 other races.
Ex ROSE RONALD	by Bay Ronald. 3rd dam of Commanderie FO, GP.
Ex WILD ROSE	by Beau Brummel.
Ex ROSE OF YORK 1880	by Speculum. Exp. France '92, dam of Roxelane FO (d. of Roi Herode) see Appx. B.
Ex ROUGE ROSE	by Thormanby, dam of Bend Or, see p. 11. Family no. 1.

Tillywhim won 2 races, she was the only produce of her dam to live. She was inbred to Isonomy at the 4th remove.

She had 12 foals of which 8 were fillies, all of them bred winners; in all she bred 9 winners of 39 races – £22,200.

TILLYWHIM'S MATINGS:

ROYAL REALM 1905 won the Gimcrack and later the Queen Alexandra Stakes. He was by Persimmon out of Sandblast by Sheen (ancestress of Espresso). He also appears in the background of another stayer, a daughter of his being 4th dam of Random Shot GC. Royal Fancy has 3 strains of Hampton, 2 at the 4th remove and 1 at the 5th.

TILLYWHIM 1912

Mated with:– ROYAL REALM

Produce:– 1918 ROYAL FANCY
won 4 races,
bred 3 wnrs.

ETHERIAL
1932 Ethnarch
won twice

ASTRID 1938
Fair Trial
won 4 races,
bred 7 wnrs.

ASTRID WOOD
1944
Bois Roussel
won twice

FEEVAGH 1951
Solar Slipper
won 3 races
bred 5 winners

FEEMOSS 1960
Ballymoss
won 3 races
bred 4 wnrs. in
UK so far.

LEVMOSS
1965 AT, GC,
Le Levanstell
His sire
*LE LEVANSTELL
1957, av. winning
dist. 7.6f; by Le
Lavandou from a
Ballyogan mare,
inbred *3 times* to
Son In Law at
4th rem.

& his sister **SWEET MIMOSA**
1967 O,
Le Levanstell.

Mated with:– TRACERY

Produce:– 1919 ENRICHMENT
won 2 small races,
bred 4 wnrs. of 13
races – £8,314

ENDOWMENT
1924
Silvern
won 5 races

BEGUM 1934
Blandford
won twice

LADY GRAND
1943
Solario
won once
bred 12 wnrs.

PHYLLIS COURT
1948
Court Martial
won once

ALCOA 1951
Alycidon
won twice

LUPE 1967 O,
Primera
won 5 races
& £53,700
Her sire
*PRIMERA 1954,
won Ebor (h'cp 14f)
exp. Japan 1969;
by My Babu from
the FO wnr. Pirette
by Deiri, inbred to
Pharos at 4th rem.
Sire also of Aunt
Edith GVI.

EMOLUMENT
1928
Gainsborough
did not win,
bred 5 wnrs.

HONORARIUM
1935
Colorado Kid
won once, bred
7 wnrs.

†GRANDMASTER
1942 Atout
Maitre
a miler, sired dam
of Sweet Solera
1,O. His sire
ATOUT MAITRE
1936 imp. UK '39
a stayer, by Vatout
from a Teddy mare,
inbred St. Simon
4 rem. Sired also
Ossian II.

Mated with:– COLORADO

Produce:– **1930 FIGARO**
A sprinting handicapper,
won 10 races. Inbred to
Cyllene at the 3rd and
4th removes; he had
furthermore a most
expensive background,
St. Simon 4th and 5th
removes, Hampton
and Isonomy 3 times
each in the 5th. He
sired the mare Edvina
(won 6 races in Ireland),
2nd dam of Martial 2
& 4th dam of *Bold
Lad 1964 (Sire of
Waterloo 1)

DINNER 1913

The origins of **DINNER** 1913. By **DINNEFORD** 1902, a miler, won the Hunt Cup at 4y carrying 7.7. By the good mid. dist. h'cpr. *Dinna Forget who ran 31 times and won 8 races, out of the otherwise undistinguished Gracie by Arbitrator.

Ex PERNELLE 1907	by PERSIMMON. Pernelle won 1 small race.
Ex NUNEATON	by BEND OR.
Ex CAPUCINE	by Camballo.
Ex MATILDA 1880	by Beauclerc, dam of Queen's Birthday and Nunthorpe. Family no. 11.

Dinner was a first foal and never won.

Not a very showy pedigree, the only classic indications being the Persimmon and Bend Or mares in the bottom line. Overleaf will be seen the effects of three very important matings.

Dinner had 10 foals, 5 of which won 14 races value £21,800.

As a result of the matings (next page) Bonne Bouche is inbred to Loved One at the 4th remove and Mellowness to St. Simon at the 2nd and 4th.

DINNER 1913

Mated with: – BUCHAN | CHAUCER | TETRATEMA

Produce: –

BUCHAN

1922 BONNE BOUCHE
did not win,
bred 4 wnrs.

GENTLEMAN'S RELISH
1926 He.
won 4½ races
bred 3 small wnrs.

MARMITE 1935
Mr. Jinks
won twice
bred 6 wnrs.

SAUCY BELLA
1941
Bellacose
won 2 sprints
bred 4 sprint wnrs.

***HARD SAUCE**
1948 Ardan
6f specialist,
sired Hard Ridden D, *Hard Tack and the dam of LASSALLE GC

MADRILENE
1951 Court Martial
won twice at 2y
bred 5 wnrs.

CAMANAE
1961 Vimy
won once (14f)

HIGH TOP
1969 2,
Derring Do

CHAUCER

1923 MELLOWNESS
did not win,
bred 3 wnrs.

BUCHANESS
1928 Buchan
did not win

BUCHANITE
1932 Son In Law
did not win,
exp. France 1937

†BREUGHEL
1939 Brantome
sprinter sired dam of Belle Ferronniere FO

FORGIVENESS
1930 Tetratema
did not win

PALOMA
1939 Bosworth
did not win

PANORAMIC
1944 Panorama
did not win
bred 4 wnrs.

HONEYMOON
Beau-Pere

†HONEY'S ALIBI
1952 Alibhai
Sired dam of Dahlia GVI (twice) (Ir.O. Wash.Int.). His sire *ALIBHAI 1938 was a well-known US stallion, by Hyperion from a Tracery mare, he sired also Traffic Judge.

MEADOW RHU
1931 Phalaris
won 3 good races

HANGING FALL
1940 Solario
won once, bred 5 wnrs.

RIDGE WOOD
1946 L, Bois Roussel
Inbred to Chaucer at 4th rem. Died young, sired nothing of note.

TETRATEMA

1928 FOUR COURSE 1,
won 4 races and £14,000.

MOLLY DESMOND 1914

The origins of **MOLLY DESMOND** 1914. By DESMOND (p. 63) out of PRETTY POLLY (p. 73), she was inbred to HERMIT at the 4th remove.

Molly Desmond won 1½ races and £1367.

She had 6 foals of which 5 won 11½ races.

The mating with SWYNFORD meant that Sarita (next page) and her sister Golden Silence were inbred to St. Simon at the 3rd and 4th removes, to Isonomy at the 4th and with a third strain of Hermit at the 4th to add to Molly's two at the 5th.

*†**PHALARIS** 1913, the good middle-distance handicapper who is today the most influential partriarch of the Bend Or dynasty and the progenitor who looks more likely to survive as a classic byword than any other. From him stems, among others, the great Nearco (p. 100). Phalaris was by the Champion Stakes winner Polymelus out of Bromus by Sainfoin, she being closely inbred to Springfield. He raced only at Newmarket during the first war, there he won 16 small races; on that straight course he did not stay beyond 9f. He was twice champion sire of winners and 3 times of broodmares, 3 of his sons have founded lines which survive today and, thanks to his grandson, Nearco, look like lasting a great deal longer. His most important products, so far as UK is concerned, were the brothers Pharos and Fairway; in USA his son Sickle was twice champion sire. Other notabilities were Pharos' sister Fairisle 1, Warden Of The Marches Ch, *†Colorado 2, E, Caerleon E, *Pharamond, Chatelaine O, =Ch, Herbalist and Bold Archer. His almost shamefully inbred (Springfield at 2nd and 3rd rems.) dam, Bromus has left no daughter to carry on her line in top class racing.

MOLLY DESMOND 1914

Pretty Polly's eldest daughter

Mated with: –

Produce: –

- **SPEARMINT**
  - **1919 SPIKE ISLAND** & his sister
    won Irish 2 & D.
    Successful stallion first in Argentina 1923, then in France and ended up in Italy. Sired Dossa Dossi (It.1.O) the 3rd dam of Psidium D (p. 74)
  - 1920 LADY MAUREEN
    did not win, exp. France 1930
    - MOIRHEEN RHUE 1928
      Swynford
      did not win
      exp. Italy '34
      - MANNOZZA
        1934 ex. Italy
        Manna
        - ***MACHERIO 1941 (exp. USA) (It. 2, L, GP Milan) Ortello**
          Champion sire Italy, sired dam of Weaver's Hall IRD. His sire ORTELLO 1926 AT, (It.D) was by Teddy from a Gorgos mare and Inbred St.Simon and Hampton 4 rem. He was imp. '57 from Italy. Sired also dam of Ossian II
    - MAURETANIA
      1930 Tetratema
      did not win
      - MONTAGNANA
        1937 Brantome
        won in Germany
        - ***OCARINA (bro.) 1947 Bubbles**
          Sired LAUSO (It. D and sire of Precipice Wood GC) ex. Canea (It. O) Lauso was imp. UK '61
        - **†GUERSANT 1949 (F2 record) Bubbles**
          won also the Hardwicke Stakes and 5 other races. Inbred Brantome 2 & 3 rems. & Tetratema 3 & 4. exp. Japan. Sired dam of Astec FD. His sire *BUBBLES 1925 won the Prix Lupin and was sired by La Farina from a Neil Gow mare, Bubbles sired also Eclair Au Chocolat AT, and Magister FD.
        - FLUTE ENCHANTEE
          1950 Cranach
          - ***LUTHIER 1965 (Lup) Klairon**
            Sire of Galiani 1975 GP
- **SWYNFORD**
  - 1924 SARITA
    won once exp. Fr. '33
    bred 2 small winners
    - SISTER SARAH
      1930 Abbot's Trace
      won twice, bred 8 winners.
      - CAERLISSA 1935
        Caerleon
        did not win, bred 8 winners.
        - EDIE KELLY
          Bois Roussel 1950
          won 1 small sprint
          bred 8 wnrs. of 24 races –£114,600
          - FIGHTING EDIE
            1956 Guersant
            exp. France 1964
            - FORMENTARA
              1968 Ribot
              - **FLYING WATER 1973 1,Ch. Habitat**
          - ***ST. PADDY 1957 D,L,E, Aureole**
            Won 7 races inc. Hardwicke Stakes. Sired Connaught E. and the dam of Dunfermline O, L
      - SYBIL'S SISTER
        1943 Nearco
        did not win, bred 7 winners.
        - SYBIL'S NIECE
          1951 Admiral's Walk.
          won 3 races at 2y; exp. Italy 1963; bred 7 wnrs. UK.
          - ***GREAT NEPHEW 1963 Honeyway**
            a miler, sired Grundy D, IRD, GVI, (Ir.2) His sire *†Honeyway 1941 Ch. & 15 other races, was by Fairway from a Papyrus mare; he sired also Honeylight 1, *Sunny Way & the dams of Pleben GP and *Diatome (Wash.Int)
      - LADY ANGELA
        1944 Hyperion
        won once, exp. Canada 1953
        - ***NEARCTIC 1954 Nearco**
          Canadian miler. Inbred Chaucer 4 rem. Sired *Northern Dancer and Nonoalco 2. see p. 98

PHALARIS

1925 MOLLY ADARE
did not win
bred 6 wnrs. of 17 races

sis. to Sarita

FEARLESS FOX
1933
Foxlaw
stayer, won Gold Vase and Gdwd. Cup. Inbred Gallinule 3 & 4 rems. Sired 2nd d. of Nelcius FD.

BRAZEN MOLLY
1940 Horus
did not win

LA PAIVA
1956 Prince Chevalier
did not win

BRIGADIER GERARD 1968
2, E, GVI, Ch. (twice)
Queen's Hussar
Inbred Fair Trial at 4th rem.
see p. 159

BRAVE EMPRESS
1939 Tiberius

ACE OF SPADES
1947 Atout Maitre

SUNSET GUN II
1955 exp. USA '55
Hyperion.
bred 3 wnrs.

STYLISH PATTERN
1961 My Babu

ARTAIUS
1974 E (3y)
(record)
Round Table

SWYNFORD

1926 GOLDEN SILENCE
won 3 races

TIP THE WINK
1934 Tetratema
won 3 good races
bred 4 wnrs.

TRIAL GROUND
1944 Fair Trial
won once, bred 5 wnrs.

PREMONITION
1950 L
Precipitation
won 8 races
Sired dam of Mon Fils 2

ORLASS 1914

The origins of **ORLASS** 1914. By ORBY p. 5.

Ex SIMON LASS 1909	by **SIMONTAULT** 1893, never ran. By St. Simon out of a mare by Trumpeter, quite well-connected though unfashionable, his fee by 1900 was only 19 gns., 10 for dams of winners. Simon Lass won 4 races, granddam of Sir Walter Raleigh (sired 2nd dam of Wolver HollowE).
Ex KILKENNY LASS	by LESTERLIN p. 15. Kilkenny Lass bred 7 winners.
Ex STANDON GIRL	by The Lambkin.
Ex STANDON	by Craig Millar, family no. 21.

Orlass, a first foal, won 5 small races. She was inbred to Galopin at the 4th remove.

She had 10 foals, 9 of them won 21½ races and £13,500.

ORLASS'S MATINGS:

*†**FRIAR MARCUS** 1912, The King's sprinter, won 9 races during the first war. Impeccably bred by the Derby winner, Cicero, from a Persimmon mare; inbred to Hampton at 4th rem. He sired Brown Betty 1, *†Beresford (sire of Portlaw), the dams of *†Dastur =Ch. (Ir.D), *†BAHRAM 2, D, L, Vigilance FO, Hypericum 1, Columcille, Kingstone and Stardust.

ORLASS 1914

Mated with: – FRIAR MARCUS

Produce: – 1923 ORISON
won twice
bred 6 wnrs.

OLIFA 1930
Buchan
won twice,
bred 9 wnrs.

STRAIGHT OFFER
1948
Straight Deal
won twice,
bred 6 wnrs.

STRAIGHT BID
1954 Solonaway
won 3 races at 2y

WINNING BID
1958
Great Captain
won 3 races

COUP DE FEU
1969 E,
White Fire III
his sire, WHITE
FIRE III 1962,
won in France,
av.7.5f., imp '65
stood in England
until exp. Japan
1969

PELEID 1970 L,
Derring Do

Mated with: – SOLARIO

1931 LOST SOUL
won twice, bred
3 wnrs. of 30 races
val. £15,800

PHASE 1939
Windsor Lad
won 3 races
bred 9 wnrs
of 33 races
– £50,300

NETHERTON & her sis. **NEASHAM** & her bro. **NARRATOR**

NETHERTON MAID 1944
Nearco
won twice,
bred 5 wnrs.
of £26,500

BRIDE ELECT
1952 Big Game
won 2 big races
bred 10 wnrs.
of 24 races –
£76,600

***HETHERSETT**
1959 L,
Hugh Lupus
fell in the Derby;
inbred to Tour-
billon at 3rd and
4th rems. Died
aged 7; sired
Blakeney D
see p. 132

NEASHAM BELLE
1948 O,
Nearco

NARRATOR
1951 Ch.
Nearco
Sired Night Off
1.

NEARLY
1940 Nearco
won 3 races
at 2y. Bred 9
wnrs. of 25 not
very big races

DIVEL 1945
William of
Valence
won once at
2y. exp.
France 1950

FLORALIE
1954
Pot O'Luck

FLOSSIE
1966 Ch.
Spy Well
SPY WELL 1960
exp. Jap. '71
US miler, by
Mark Ye Well
ex a mare by
Spy Song

CLAUDETTE
1949 Chanteur
II
won once at 2y,
bred 5 wnrs.

SIPSEY BRIDGE
1956 Abernant
exp. Fr. '62
won twice

***DERRING DO**
1961 Darius
Inbred Nearco
3 & 4 rems. &
Pharos 4th.
Sired High Top
2 and Peleid L,
(v.s.)

a sis. to Lost Soul

1933
VENTURE-
SOME
won once
bred 4 wnrs.

DARING MISS
1939
Felicitation
won 3 small
races bred 8
wnrs. of 28
races –
£29,800.

FLATTER
1947
Rockefella
did not win
bred 5 wnrs.

FLATTERING
1961
Abernant
won twice at
2y.

HUMBLE
DUTY 1967
1, Sovereign
Path
won 8 races
and £62,100
Inbred to
Hyperion at
4th rem.

SEAWEED 1916

The origins of **SEAWEED** 1916. By SPEARMINT p. 73, at the time of this mating had no reputation as a sire.

Ex SEADUNE 1908	by AYRSHIRE. Seadune won 2 small races, bred 3 winners. Her other daughter, **Sea-Moss** (exp. France '21) by William The Third, was granddam of Peniche FO and 3rd dam of *†Prince Bio (Sire of Sicambre), p.162, App. D.
Ex SEA-DOWN by	**ORVIETO** 1888 (bro. to Laveno Ch. p. 25), a useful middle distance horse, ran 29 times and won 13, sired Picton. Sea-down bred 8 winners of 26 races.
Ex NEW ZEALAND	by Isonomy.
Ex AUSTRALIA	by Hermit; over 100 years back down this line to the nearest classic connection; the supporting sires all the way are of good repute, why no classic wins or contributors? More important still, why the blossoming which seems to start with Seadune? I cannot answer other than to say feebly that the science of breeding, termed proverbially inexact, never looked more so.

Seaweed herself never won and was exported to France 1924. Family no. 1.

She bred only the two winners above, she had 6 foals.

SEAWEED'S MATINGS:

***†BRULEUR** 1910 GP, a great stayer, by †Chouberski from Basse Terre by †Omnium II FD, his pedigree contained no St. Simon at all. In addition to the two above, he sired *†Ksar FD, AT (twice), his successor in the French dynasty of Gardefeu (see p. 178), Pot au Feu FD, Priori AT, Madrigal FD, Samos AT, Palais Royal and the great stayer Finglas. He sired also the dam of Lysistrata FO.

SEAWEED 1916

Mated with: – BRULEUR

Produce: –

1926 HOTWEED FD, GP, (Lupin) To stud first in France where he got Pearlweed 1932 FD (no good at stud either in UK or **Argentina) and** his sister Pearl Drop, the 2nd dam of *†Tourment and Flocon E. Hotweed came '36 to England, with little success.

and his sister

1928 BRULETTE O imp. UK 1932 won 7 races in UK & France including the Goodwood Cup, Jockey Club Stakes and the Cadran. She bred 8 moderate winners.

TROPICAL SUN 1940 Hyperion won 3 races

OPEN COURT 1952 Court Martial

COURTING 1961 Quorum (see p.160)

LAXMI 1966 Palestine

ENSTONE SPARK 1975 1 Sparkler see Ap. A.

BELLE SAUVAGE 1949 Big Game won once, bred 6 wnrs.

NOBLE LASSIE 1956 Nearco won 2 good races

***VAGUELY NOBLE 1965 AT, Vienna** Inbred Hyperion 3 & 4 rems & to Bahram 4th. Sired Dahlia GVI (twice) (Ir. O, Wash. Int) and Empery 1973 D, Exceller 1973 GP. (see p. 173)

PEARL MAIDEN 1918

The origins of **PEARL MAIDEN** 1918. By †PHALERON p. 14.

Ex SEASHELL 1908	by ORME. Seashell won twice.
Ex RYDAL FELL	by LADAS.
Ex RYDAL	by BEND OR. Granddam of Troutbeck L.
Ex WINDERMERE 1870	by MACARONI. Dam of *†Kendal (exp. S. America 1901) and *†Muncaster. Family no. 16.

Pearl Maiden did not run, she was exported finally to France after frequent visits. She has a tremendous pedigree, inbred to both Galopin and Bend Or at the 4th remove.

She bred 10 winners.

PEARL MAIDEN'S MATINGS:

TETRAMETER 1917 raced 4 seasons and won 6 sprints. He was by The Tetrarch from a mare by Love Wisely; he sired also the dam of Loliondo, otherwise nothing of much note.

***†LE CAPUCIN** 1920 FD, by Nimbus (1910) out of a mare by Sidus (by St. Simon). In addition to the distinctions above he sired the dam of Sun Cap O.

***†FOXHUNTER** 1929 GC, by Foxlaw (p. 80) from Trimestral (dam of Trimdon GC (twice)) by William The Third. Certainly bred to stay. He was sent to France in 1933 where he did quite well at stud, returned to UK for the war and exported to Argentina in 1945, there he did well too. Sired the dam *Fast Fox.

PEARL MAIDEN 1918

Settled in France after frequent visits from UK

Mated with: – TETRAMETER
- Produce – 1924 MUCI exp. Italy '26 won 6 races there
  - †**MUZIO 1933 Ksar** Italian, inbred Omnium II 3 & 4 rems. Sired dam of Nuccio AT, (It. D)
  - MURCIA 1940 Pilade
    - MAGGIOLINA 1946 Nakamuro
      - **MOLVEDO 1958 AT, Ribot** Inbred Pharos 4 rem. At stud in Italy

LE CAPUCIN
- **1928 PEARL CAP FO, AT,** Inbred St. Simon 4 rem.
  - **PEARL DIVER 1944 D, imp. UK '48 Vatellor** exp. Japan '57
  - SEED PEARL 1950 Tourment did not win
    - **FINE PEARL 1963 FO, Fine Top** (p. 46)

HOTWEED
- **PEARL WEED** and his sister **1932 FD** imp. UK 1936 from France, then exp. **Argentina '44.**
- PEARL DROP 1933 did not win
  - FRAGMENT 1938 Shred won 1 good race.
    - *†**TOURMENT 1944 (F2, FL) Tourbillon** Inbred Bruleur 3 & 4 rems. Sired Cerisoles FO and the dams of *Sanctus II FD, GP & Fine Pearl FO.
    - **FLOCON 1946 E, Fastnet** imp. UK 1951 exp. Holland 1958

FOXHUNTER
- 1936 SILVER FOX II imp. UK '37. Inbred St. Simon 4 rem. did not win, bred 3 wnrs.
  - DONAH 1942 Donatello II won twice at 2y bred 5 wnrs.
    - OLD DUTCH 1959 Fastnet Rock won twice bred 4 wnrs. to date
      - **SLEEPING PARTNER 1966 O, Parthia**

ZARIBA 1919

The origins of **ZARIBA** 1919. By SARDANAPALE p. 85.

Ex ST. LUCRE by ST. SERF, see Dame Masham p. 11.

Zariba won 13 races and was 2nd in the French Oaks. She was inbred to St. Simon at the 3rd and 4th removes. Family no. 9.

Her 4 winners netted £18,800 in England and Ireland alone.

ZARIBA'S MATINGS:

*†**POMMERN** 1912 2, D, L, all his victories took place at Newmarket in 1915 during the first war; nevertheless he was easily the best of his age. By Polymelus from a mare by St. Hilaire (by St. Simon ex Distant Shore by Hermit), he was inbred to Hampton at the 3rd remove and to the mare Distant Shore at the 4th. His career at stud did not match his ability as a racehorse, however, he got Adam's Apple 2 and the dams of Campanula 1 and †Shred. L'Esperance is inbred to Hampton and St. Simon at the 4th remove.

ZARIBA 1919

Whose dam came from UK to France in 1908

Mated with: – POMMERN; GAINSBOROUGH

Produce: –

- 1925 L'ESPERANCE (by POMMERN)
  bred 4 winners
  - FLOWER 1936
    Pharos
    - NAFAM
      Abjer
      - MISS FRANCE
        1946 Jock II
        exp. UK 1947
        did not win
        bred 7 wnrs.
        - **FIDALGO**
          1956 (Ir. D)
          Arctic Star
          Inbred Pharos and Asterus 3 & 4 rems. Exp. Japan 1966. Sired dam of Full Dress II 1. His sire ARCTIC STAR 1942 never ran; by Nearco from a mare by Winalot; sired also Arctic Storm Ch.
  - SOURYVA 1930
    Gainsborough
    won twice flat and 4 times hurdles
    bred 4 wnrs.
    - PHARYVA 1936
      Pharos
      did not win
      bred 7 other wnrs.
      inc. an (F1)
      - **GALCADOR**
        1947 D, Djebel
        Inbred Bayardo 4 rem. exp. Japan '59
    - THEANO
      1940 Tourbillon
      won twice, bred 5 wnrs.
      - **PHILIUS II**
        1953 FD,
        Pharis
      - ASTANA
        Arbar
        won 1 good race
        - **CREPELLANA**
          1964 FO,
          Crepello
          Inbred Tourbillon 3 & 4 rems.
- **1928 GOYESCAS Ch.** (by GAINSBOROUGH)
  also won Hardwicke Stakes; inbred St. Simon 4 rem.
  died in training.

CORONACH
|
1932 CORRIDA AT (twice)
|
***†COARAZE 1942 FD, Tourbillon**
Exp. Brazil 1954 Sired *EMERSON and the dams of Tambourine II IRD, White Label GP and Nasram GVI.

ASTERUS
|
1933 *ABJER
A good 2yo in England, to stud in France, sired Caravelle FO, *Tifinar, *†Nosca and Hierocles. Abjer died aged 6.

TOURBILLON
|
1934 GOYA II
Won Gimcrack & St. James' Palace Stakes, 2nd in 2000 and Ch. Sired Sandjar FD, Corteira FO, (F1), Good Luck FD, Asmena O and Goyama, also the dams of Tahiti FO and *Hugh Lupus Ch. (Ir. 2) Goya II was exp. to USA in 1947 having been champion sire in France once.

MUMTAZ MAHAL 1921

The origins of **MUNTAZ MAHAL** 1921. By The Tetrarch p. 93.

Ex LADY JOSEPHINE by Sundridge, see p. 11.

Mumtaz Mahal, the "Flying Filly", won 7 races and £13,900. In spite of 6f being really the limit of her distance she was 2nd in the 1000 gns.

She had 8 foals, 5 of which won 14½ races and £12,900.

MUMTAZ MAHAL'S MATINGS:

***†BLENHEIM** 1927 D, won 4 races at 2y, injured himself after the Derby and did not run again. He went to stud in France in 1932 and was sent to USA in 1936. He made a great name for himself in both continents, in France siring Mumtaz Begum and Mahmoud (next page) and then, most important of all, Donatello II who won everything in Italy and ended up in Newmarket, a most powerful link in the Isonomy dynasty. By Blandford from Malva (dam of *King Salmon and †His Grace) by Charles O'Malley, he was inbred to Isinglass at the 4th remove. His daughters bred also Le Paillon AT, Argur E, Tabriz, Hill Gail and Mark Ye Well. See App. C.

MUMTAZ MAHAL 1921

Finally exported France 1937

Mated with: – GAINSBOROUGH

Produce: – 1928 MAH MAHAL exp. Fr. '31 won 1½ small races bred 6 wnrs. of 15 races val. £21,000

- ***†MAHMOUD fld. Fr. imp. UK '34 1933 D, (record) Blenheim** First foal. Won 4 races, 2nd in 2000 and 3rd in Leger. To stud aged 4, sold to USA 1940 where he became champion sire in 1947. In the European context of this book he appears so far as the sire of †Mr. Trouble (sire of dam of Sir Ivor), the maternal grandsire of Masaka O, (Ir.O) & †Fastnet Rock.
- MAH IRAN 1939 exp. USA '55 Bahram won 5 races bred 6 wnrs.
  - **MIGOLI 1944 exp. USA '59 E,Ch,AT, Bois Roussel** won 8 other big races and ran 2nd in the Derby. His results at stud, despite good mares, have been disappointing.
  - and his sis. STAR OF IRAN 1949 Bois Roussel won once, exp. USA 1956
    - **PETITE ETOILE 1956 Petition 1,O,Ch.** won 14 races & £67,700 inbred Gainsborough 4 rem.

Mated with: – BLANDFORD

Produce: – **1931 †BADRUDDIN** Won 4 races of 1 mile as a 3yo. Went to stud in France first, then exp. Argentina. Sired Loliondo, but far more important, also the mare Perfume II, dam of *†My Babu 2 and *SAYANI (won Cambridgeshire car. 9.4 at 3y and became champion sire in France; exp. Brazil 1953).

Mated with: – BLENHEIM

Produce: – 1932 MUMTAZ BEGUM foaled in France, imp. UK 1934 won 2 small races, bred 8 wnrs. of 22 races – £11,400

- SUN PRINCESS 1937 Solario did not win bred 9 wnrs. of £19,300
  - ***ROYAL CHARGER 1942 Nearco** a 6f specialist, exp. USA 1953 having sired Happy Laughter 1, Gilles De Retz 2, *Turn To and Ossian II. see p. 153
- ***†NASRULLAH 1940 Ch. Nearco** won 4 other races including the Coventry Stakes run at Newmarket. He was very temperamental, a fault which he was inclined to pass on. He stood in UK before going to USA in 1950, where he was certainly the most influential of Nearco's American dynasty. While in UK he sired the winners of 7 classics including the stayer *NEVER SAY DIE D, L, (from an American mare) also the important stallions *†Grey Sovereign and *Princely Gift. In the States, whence four of his sons have so far come into the sphere of European racing, his first crop produced Nashua (won 22 out of 30 starts in 3 seasons) who sired the dam of Roberto D, then Bold Ruler, sire of *Bold Lad 1962; Bald Eagle (Wash.Int. twice) sire of San San AT and Never Bend sire of Mill Reef D, **GVI, E, AT.** **see p. 152**
- MALINDI 1947 Nearco won twice exp. France 1952
  - ***PRINCE TAJ 1954 Prince Bio** a miler, ran 2 seasons in France then another two in USA. Went back to France to stud 1960. Sired Astec FD, and the dam of Tennyson GP.
- DODOMA 1939 Dastur
  - OMELIA 1950 Owen Tudor exp. USA 1957
    - KAKIA 1956 Sunny Boy III
      - ZAKYNA 1965 Charlottesville did not win
        - **FUNNY HOBBY 1974 GP Meadow Mint** Meadow Mint 1969 was a good 2 yo (won at Ascot) but deteriorated at 3y winning only one moderate 9f race. He died very soon after going to stud in 1973

Mated with: – RUSTOM PASHA

Produce: – 1934 RUSTOM MAHAL did not run, bred 3 wnrs. of 22 races. val. £28,800

- ***†ABERNANT 1946 Owen Tudor** won 14 races and £26,300; 2nd by a short head in Nimbus's 2000; His max. winning distance was 7f, his most usual 5f. He sired Abermaid 1 and the dams of Caergwrlie 1, Humble Duty 1 and *Derring Do.

DORINA 1923 FO

The origins of **DORINA** 1923 FO. By LA FARINA p. 101.

Ex DORA AGNES 1915	by *†**ROI HERODE** 1904, a staying handicapper imported from France to UK in 1907. By Le Samaritain from Roxelane by War Dance. Sired also *†The Tetrarch, Heredote, Chrysolaus and the dams of Strip The Willow FD, GP, Zionist and American Flag (Belmont). Roi Herode was inbred to Thormanby at the 4th remove. Dora Agnes was exp. France 1920.
Ex D'AMOUR	by Raconteur.
Ex AMORA	by Amphion.
Ex MEDORA 1890	by Bend Or and 4 generations back to **Little Agnes** 1856, the first foal of Miss Agnes (p. 7) and founder of 3 classic families. Family no. 16. Medora was dam of Zinfandel GC.

Dorina, in addition to the French Oaks, won 5 other races. She was inbred to St. Simon at the 4th remove. Dorina had 6 foals in France.

DORINA 1923 FO

(Her dam came from UK to France in 1920)

Mated with:– ASTERUS

Produce:–

1931 *† ASTROPHEL raced 4 seasons all over France and sometimes in Algiers, won 20 races over an av. of 10.5f. Sired Bagheera FO, GP and the dam of Puissant Chef AT (FL).

and his sister

DULCE II 1932 imp. UK 1937 from France won 3 races bred 5 wnrs.

- RED SUNSET 1941 Solario won once bred 9 wnrs.
  - CREPUSCULE 1948 Mieuxce won once
    - **HONEYLIGHT 1953 1, Honeyway** won 5 races
    - ***† CREPELLO 1954 2, D, Donatello II** has so far sired *Busted E,GVI, Caergwrlie 1, Crepellana FO & Mysterious 1,O; his daughters have bred Altesse Royale 1,O, and Genevra O.
    - **TWILIGHT ALLEY 1959 GC, Alycidon** Inbred Gainsborough 4 rem.
  - ***SUNNY WAY 1957 Honeyway** won 5 races up to 16f. exp. Venezuela 1971; sired Scottish Rifle E,
- MEHMANY 1943 Mieuxce won 5 races
  - ***† SHIKAMPUR 1950 exp. Fr. '53 Tehran** won 4 middle distance races, sired the dam of Niksar 2.

PRIDE OF HAINAULT 1924

The origins of **PRIDE OF HAINAULT** 1924. By ***HAINAULT** 1914, an injury interfered with his racing career, nevertheless he won 6 small races; he went to stud in Ireland and died after 6 seasons. By Swynford out of **Bromus** by Sainfoin, the dam of Phalaris, he could not have been better bred. He sired also the dam of Wychwood Abbot Ch. (twice) and Clang, the granddam of Nimbus 2, D.

Ex MARTAGON'S PRIDE 1917 Won twice	by ***LOMOND** 1909, won Gimcrack and 5 other good races 7-10f, stood first in Ireland then exp. Argentina.
Ex LADY GRAVEAIRS	by Martagon.
Ex STE. NITOUCHE	by St. Simon.
Ex FLEUR DE MARIE	by Hermit, Family no. 8.

Pride Of Hainault never ran, she was exported to France 1929. She was inbred to St. Simon 3 times at the 4th remove.

She had 11 foals, she bred 7 winners in France.

PRIDE OF HAINAULT'S MATINGS:

***†TEDDY** 1913 FD, raced during the first war; apart from the substitute FD, he won 2 important races for stayers in Spain. He stood in France until sent to USA in 1931 aged 18; he was a success there too before he died aged 23. By the FD & GP winner Ajax (by Flying Fox) from the mare **Rondeau** (had just had 3 dead foals in succession) by Bay Ronald, he had two strains of Bend Or and 3 of Galopin between the 3rd and 5th remove. His most important offspring were *†Asterus Ch. (F2) (p.86, 148, App. D), the brothers *†Sir Galahad III (F2) & *Bull Dog – both of them champion sires in USA – Ortello AT, (It.D) and †Shred. However, it was his son Aethelstan, from a humble mare by a sire of jumpers, who kept Teddy's place for him in the Bend Or dynasty. His daughters bred the winners of 5 European classics, a 'Belmont' and a 'Preakness' in The States, *†Vatellor, †Amfortas, †Norseman and 3 other useful sires. His mating with Pride of Hainault provides Mercia and Philippa Of Hainault (next page) with no less than 8 strains of Galopin in their background, see p.29, 148, 161, 172.

PRIDE OF HAINAULT 1924

(Arrived in France 1929)

Mated with: – TEDDY

Produce: –

1931 MERCIA
won once at 2y
bred 5 wnrs.

LE PETIT PRINCE
1951 FD, Prince Bio
FD was won in a storm, he broke down afterwards, sent to Argentina.

and her sister

1932 PHILIPPA OF HAINAULT
won 4 races inc. the Free h'cp.
Bred 8 wnrs.

L'ORGUEILLEUSE
1947 Vatellor

GREAT SUCCESS
1950 Niccolo dell'Arca

PARDALLO
1963 GC, Pardal
Inbred Teddy
4 rem.

and her sister

PHILIPPA II
exp. France 1950
1943 Vatellor

LA PERIE
1951
Hyperion

BELLE SICAMBRE
1961 FO, Sicambre

***†PHIL DRAKE**
1952 D, GP, Admiral Drake
Inbred Teddy
3 & 4 rems.
Died March 1963.
Sired dam of *BON MOT III AT.

FANCY FREE 1924

The origins of **FANCY FREE** 1924. By †STEFAN THE GREAT p. 10.

Ex CELIBA 1916	by BACHELOR'S DOUBLE p. 79. Celiba did not run, bred 6 winners.
Ex SANTA MAURA 1893	by ST. SIMON.
Ex PALMFLOWER 1874	by THE PALMER. Palmflower won 1 good race, dam of St. Florian and Musa O (see Appx. B), granddam of Symington, Mirska O and Montmartin FD, GP gt. granddam of Buckwheat.
Ex JENNY DIVER 1866	by Buccaneer, from whom sprang two other classic branches. Family no. 20.

Fancy Free had 9 foals, 8 of which won 16 races and £44,300.

HER MATINGS:

*†**APELLE** 1923 (It.D) was foaled in Italy, came to England in 1926, won the Coronation Cup and 4 more races during the next 2 seasons and then went to stud. His first crop produced Cappiello GP, (Lupin) which promoted him to champion sire in France; he sired also the dams of †Straight Deal D, *Tenerani and Aquino II GC. Sired by Sardanapale from a St. Frusquin mare (ex a Cyllene mare), he was inbred to St. Simon at the 3rd and 4th removes.

FANCY FREE 1924

Mated with: – APELLE

Produce: –

1932 SPRINGTIME
won 1 small race
bred 11 wnrs.

NEBERNA 1942
Nearco
won once
bred 4 wnrs.

CHORUS BEAUTY
1952 Chanteur II
won twice

WINDMILL GIRL
1961 Hornbeam
won 2 good races
2nd in Oaks
bred 4 wnrs.

BLAKENEY 1966
D, Hethersett
Sired so far Juliette Marny O and Julio Mariner L. Blakeney is inbred Nearco at 4th rem.

MORSTON
1970 D, fld
Fr. imp. '71
Ragusa

Mated with: – FAIRWAY

1936 *†BLUE PETER
2,D,E,
Won also the Blue Riband Trial Stakes at Epsom – all in the year that war broke out. The Leger was cancelled. He did not race again after the Eclipse and went to stud at 4y but, possibly due to the war, he was not the success at stud that he was on the racecourse. Perhaps Ocean Swell D,GC was his best son who sired †Fastnet Rock, but little else of consequence; his other sons were Peter Flower Ch., Botticelli GC and the unfortunate Blue Train who won 3 good races, went unsound and then became unfertile. Blue Peter sired also the dams of *King Of The Tudors E, Hermieres FO, *Hard Tack and *Princely Gift.
Blue Peter was inbred to St. Simon at the 4th remove.

ROSE RED 1924

The origins of **ROSE RED** 1924. By SWYNFORD p. 36.

Ex MARCHETTA 1907	by MARCO p. 39. Marchetta won 3 races, bred 8 winners.
Ex HETTIE SORREL	by Peter. Hettie Sorrel bred 6 winners.
Ex VENUS'S LOOKING GLASS	by Speculum.
Ex SYRINGA 1878	by King Lud, granddaughter of the famous Queen Bertha 1860 O, dam of the winners of 4 classics and granddam of the winners of 3. Family no. 1.

Rose Red won two races. She was inbred to Hermit 3 times at the 4th remove.

She had one other foal, by Fairway, which won 3 good races.

ROSE RED'S MATINGS:

*†**WINALOT** 1921, a middle distance handicapper, won 7 good races. By Son In Law from the prolific mare Gallenza (7 winners of 25½ races – £14,300) by Gallinule, he sired *†Coup De Lyon (won 5 races and became a good sub-classic sire in Ireland) and the dams of Arctic Star, Belle Of All 1, and High Peak.

*†**KANTAR** 1925 AT (3y), 2nd FD and 2nd AT next year, by *†**Alcantara II** FD (by Perth ex a mare by Le Sancy) from a mare by Chouberski, his pedigree contained 3 strong lines of Herod – to wit, Chouberski, Le Sancy and Omnium II which have occurred in one pedigree so rarely that I must draw attention to them. It could only happen in France where all the main Herod lines were extant. He sired †Victrix (FL) and the dams of Caravelle FO, (F1) and *†Klairon (F2). Kantar was exp. USA in 1936.

ROSE RED 1924

Mated with:– WINALOT

Produce:– 1929 GODETIA
won 2 small races

HEAVENLY WIND
1937 Tai Yang
did not run,
bred 6 wnrs. of
18 races.

MONSOON 1941
exp. USA '55
Umidwar
won twice at 2y
bred 4 wnrs.

FESTOON 1951 1,
Fair Trial

Mated with:– KANTAR

Produce:– 1931 ROSETTA
did not win,
was foaled in France
bred 6 wnrs.

KEYSTONE 1938
Umidwar
won well, bred
5 wnrs.

KEYBOARD 1943
Bois Roussel
did not win

BLUE PRELUDE
1951
Blue Peter
won 2 good races,
bred 4 wnrs.

BLEU AZUR
1959 Crepello
won twice

ALTESSE ROYALE
1968 1, O,
St. Crespin III
Inbred Donatello II
at 3rd and 4th rem.

STONE OF FORTUNE
1939 Mahmoud
did not win

***FASTNET ROCK**
1947
Ocean Swell
Foaled in France
won 22 races,
max 10f. Exp.
Sweden 1961. Sired
dams of Roseliere FO,
and Sleeping Partner
O.

and her sister

MOONSTONE
1940
Mahmoud
won once, bred
9 wnrs.

TWILIGHT HOUR
1950 Nearco
won twice
bred 5 wnrs.

NIGHT ATTIRE
1966 Shantung
did not win

NOCTURNAL
SPREE 1972 1,
Supreme Sovereign
Her sire SUPREME
SOVEREIGN 1964,
foaled in France,
imp. UK 1975; won
4 good races, av. 7.1f.
By Sovereign Path
from a Vatellor mare,
inbred Nearco at 4th
rem.

PAVING STONE
1946 Fairway
won once, bred
5 wnrs.

SHIPYARD
1963 Doutelle
won 3 races

BUSTINO 1971 L,
Busted
won 5 races inc.
Coronation Cup; 2nd in
epic race for GVI, to
Grundy by a neck.

HYPERION

|

1936 AURORA
won once
bred 10 wnrs. of
over £51,000

BOREALIS 1941 **exp. Swe. '61** **Brumeux** won 7 races inc. Coronation Cup, 2nd in Leger. Sired dam of Highclere 1, FO, also the 2nd d. of Royal Palace 2,D,E, GVI. His sire BRUMEUX 1925 was a stayer by Teddy from an Alcantara II mare	AMBOYNA 1943 Bois Roussel did not win, exp. France 1951 | ASH PLANT 1948 Nepenthe | ***CELTIC ASH** **1957 (Belmont)** **Sicambre** **exp. USA 1958** **to Jap. '70** Sired Athens Wood L.	and her sister	WOODLARK 1944 exp. USA '56 Bois Roussel won once at 2y bred 4 wnrs. | SKYLARKING 1950 Precipitation won 3 races | **LARKSPUR 1959** **D, Never Say Die** Inbred Vatout at 4th rem. Exp. Japan 1967	***† ALYCIDON** **1945 GC,** **Donatello II** Inbred Swynford 3 & 4 rems. Won also Gdwd. & Don. Cups & Jockey Club Stakes, 2nd in Leger. His progeny stayed but were not plodders. Sired Meld 1,O,L, *Alcide L,GVI, Twilight Alley GC, Home- ward Bound O & *Kalydon. His daughters bred Never Too Late 1,O, Parbury GC, Lupe O, and Charlottown D, (ex Meld).	and his brother	**† ACROPOLIS 1952** **Donatello II** won 7 races of up to 13f including the Great Voltigeur Stakes, a better sire of jumpers than of flat racers, he sired the dam of Athens Wood L.

LA TROIENNE 1926

The origins of **LA TROIENNE** 1926. By *†TEDDY FD p. 129.

Ex HELENE DE TROIE 1916	by †**HELICON** 1908, won 3 middle distance races. By Cyllene from **Vain Duchess** (3rd dam of Manna 2, D and Sandwich L) by Isinglass. Helicon also sired the dam of *†Rialto. Helene de Troie was dam of Adargatis FO (p. 21).
Ex LADY OF PEDIGREE	by †**ST. DENIS** p. 8. Exp. France 1914.
Ex DOXA	by MELTON.
Ex PARADOXICAL	p. 19. Family No. 1.

La Troienne was foaled in France and exported to USA in 1931.

She had 14 foals and bred 10 winners, she lived to be 28.

LA TROIENNE'S MATINGS:

***†BLACK TONEY** 1911, by Peter Pan from a Ben Brush mare, his granddam (by Galopin) having been exported from UK. He sired Black Servant, Balladier, Bimelech (Belmont & Preakness) and the dam of *†Relic.

BUBBLING OVER 1923, won Kentucky Derby; by North Star III (by Sunstar from a St. Angelo mare) ex a Sweep mare going back 4 generations to Atalanta by Galopin (dam of Ayrshire). He sired Burgoo King (Ken.D. & Preak).

†BLUE LARKSPUR 1926 a miler, by Black Servant from a mare by North Star III. Sired Blue Swords (sire of dam *†Hail To Reason), the dam of †Bull Page (sire of dam of Nijinsky 2,D,L,IRD,GVI) and of the 3rd dams of *Never Bend and Roberto D. So his influence in European racing has been very great.

LA TROIENNE 1926

Whose dam came from UK to France. In USA since 1931

Mated with: – BLACK TONEY

Produce: –

- BIG HURRY won 4 races, bred 10 wnrs.
  - †**GREAT CAPTAIN 1949 imp. '56 War Admiral** A stayer, sired dam of Peleid L & Coup de Feu E.
  - & his sister inbred Ben Brush 4 rem.
  - SEARCHING 1952 War Admiral won 25 races.
    - PRICELESS GEM 1963 Hail to Reason won 7 races
      - **ALLEZ FRANCE 1970 FO, AT, (F1) Seabird II** ran 21 times, won 13 races including the Prix Ganay, she was 2nd in the AT at 3y. Her winnings exceeded £490,000.
- and her bro. **1937 BIMELECH** won 11 races inc. Belmont and Preakness. Sired granddam of *Never Bend.

Mated with: – BUBBLING OVER

Produce: –

- BABY LEAGUE won once at 2y, bred 9 wnrs.
  - STRIKING 1947 War Admiral won 3 races
    - GLAMOUR 1953 Nasrullah won 6 races, bred 8 wnrs.
      - **BOUCHER 1969 L, exp. Australia '74 Ribot** Inbred Pharos 4th rem.

Mated with: – BLUE LARKSPUR

Produce: –

- 1947 BELLE OF TROY imp. UK 1954
  - ***IMMORTALITY 1956 Never Say Die** Never ran, exp. Argentina 1967, Sired FLEET 1.
- & her sister BELLE HISTOIRE did not win, bred 8 wnrs.
  - †**ROYAL RECORD II 1958 Nasrullah** won 12 middle dist. races in US. Imp. UK 1963, exp. Japan 1968. Sired dam of Polygamy O.

DOUBLE LIFE 1926

The origins of **DOUBLE LIFE** 1926. By BACHELOR'S DOUBLE p. 79.

Ex ST. JOAN 1918	by **WILLBROOK** 1911, won 11 races including the Doncaster Cup; by Grebe (a Bend Or horse with little other distinction) from a mare by the equally unimportant Royal Emperor. Willbrook died after only a year at stud. St. Joan won 2 small races bred 4 winners of 18 races.
Ex FLO DESMOND	by DESMOND. Flo Desmond bred 4 winners.
Ex FLIGHTY FLO	by Flying Hackle.
Ex **a chestnut mare 1897**	**by Timothy** and 6 generations back to the great key mare Crucifix 2, 1, O. Family no. 2. This mare was dam of Righ Mor and ancestress of Bellacose.

Double Life had 8 foals to live, 4 of them won.

DOUBLE LIFE'S MATINGS:

***†HURRY ON** 1913 L, a huge 17 h.h. horse, unbeaten in 6 races from 8 to 18f, ran only at 3 years. He was champion sire at a time when there was Phalaris, Gainsborough and Son In Law to compete with. By the moderate Marcovil (won the Cambridgeshire aged 5 car. 7.11) from Toute Suite by Sainfoin, he was inbred to Hermit at the 4th remove. There is no trace of St. Simon anywhere in his pedigree. The Matchem dynasty had been declining for some years as far as UK was concerned; it was Hurry On who revived it. His first crop produced the Derby winner Captain Cuttle (exp. Italy and died aged 16), 2 more Derby winners came later including the great Coronach (see p. 100); in all, 10 classic races fell to his offspring but it was not until he was 20 that Precipitation came along.

Hurry On's daughters bred the winners of 7 classics, among them *†Court Martial 2, Ch.

DOUBLE LIFE 1926

Mated with: – HURRY ON

Produce: –

1933 *†PRECIPITATION GC, won also 6 other races, av. 12.6f He has succeeded in maintaining the Matchem dynasty in UK through his son Sheshoon who has a chance of saving the line from extinction over here. Man O'War, in this respect, was the chief hope in USA, having got War Admiral and War Relic; the latter's grandson, Pieces of Eight E, has recently gone to stud in UK. I guess that the American branch has a better chance of survival than ours. Precipitation sired 7 classic winners and his daughter Amber Flash (Jockey Club Cup) was dam of Ambiguity O, and another daughter bred Larkspur D. see p. 135

Mated with: – BAHRAM

Produce: –

1938 DOUBLETON did not win bred 5 wnrs. of 12 races.

DAILY DOUBLE 1943 Fair Trial won 4 races, bred 5 wnrs.

MELD 1952 1, O, L, Alycidon Inbred Blandford at 4th rem. won all told 5 races

CHARLOTTOWN 1963 D, exp. Australia 1976. Charlottesville

and her bro.

1940 *† PERSIAN GULF did not race at 2y, won nothing at 3 but won the Coronation Cup and 3 other races as a four-year-old. He sired Zabara 1, Zarathustra GC, (Ir. D, L) and Parthia D. A daughter of his bred Intermezzo L.

FEOLA 1933

The origins of **FEOLA** 1933. by FRIAR MARCUS p. 115.

Ex ALOE by Son In Law, see ALTOVISCAR p. 79. Family no. 2.

Feola won twice.

She had 10 living foals.

FEOLA'S MATINGS:

***†SIR COSMO** 1926, won the Free Handicap, just got a mile. By the sprinter The Boss (by Orby) from a Desmond mare. Sired the very fast Bellacose and *†Panorama.

***†KING SALMON** 1930 E, won in all 4 races including the Coronation Cup, ran 2nd in the 2000 and Derby. By Salmon Trout L from Malva (dam of Blenheim) by Charles O'Malley. He stood in UK for 9 seasons and was then exp. to Brazil in 1943.

***†DONATELLO II** 1934, see PRETTY POLLY p.32, 73-75, 169, 174, App. D.

†STRAIGHT DEAL 1940 D, won 5 races at 2 and 3 years and went to stud at 4, he moved to Ireland in 1953; his success was derived from two daughters, Above Board (next page) and Double Deal II, the dam of *Silly Season Ch. By Solario from an Apelle mare of a line remarkable for 3 generations of infertility though going back to WE SHALL REMEMBER, he was inbred to St. Frusquin at the 4th remove.

FEOLA 1933

Mated with: –	UMIDWAR	SIR COSMO	KING SALMON	HYPERION	DONATELLO II	STRAIGHT DEAL
Produce: –	1938 FORETASTE won 1 small race bred 2 wnrs. UK exp. France 1955	1941 KNIGHT'S DAUGHTER won 3 races exp. USA 1952	**1942 KINGSTONE** won 9 races, Sired dam of Precipice Wood GC.	**1943 HYPERICUM 1.** won 2 other races, bred 6 winners.	1945 ANGELOLA won 4 good races (12 to 14f), ran 2nd in Oaks. Bred 4 wnrs. – £38,100	1947 ABOVE BOARD won 2 good races bred 4 wnrs – £20,800.
	WINDY CLIFF 1955 Hard Sauce foaled in France	***ROUND TABLE 1954 (Am.D) Princequillo** won 43 races sired *Baldric II 2, Ch.		HIGHLIGHT 1956 Borealis won two 12f races, bred 5 wnrs. so far	***AUREOLE 1950 GVI Hyperion.** Won altogether 7 races inc. the Coronation Cup and the Hardwicke Stakes. On going to stud in 1955 he was immediately booked up for 3 years, and justifiably, for his first crop yielded *Saint Crespin III E, AT (sire of Altesse Royale 1, O,) now in Japan and his 2nd *St. Paddy D,L,E and *Vienna, both showing signs of establishing Aureole as the founder of a dynasty of his own. He sired also Aurelius L, *Miralgo (sire of Roll of Honour GP and now in Japan) and Provoke L.	
	LASSALLE 1969 GC, Bon Mot III			**HIGHCLERE 1971 1, FO, Queen's Hussar** Inbred Hyperion at 3rd rem. & Fair Trial at 4th.		
					***DOUTELLE 1954 Prince Chevalier.** Won 7 races and £10,400, died young, sired Canisbay E and the dam of Bustino L.	**†ABOVE SUSPICION 1956 Court Martial.** Won the St. James' Palace Stakes and the Gordon Stakes at Goodwood of nearly 1½ miles. He was inbred to both Son In Law and Gainsborough at the 4th remove. He sired the dam of Hard to Beat FD.

SCHIAPARELLI 1935

The origins of **SCHIAPARELLI** 1935. by SCHIAVONI (p. 83).

Ex AILEEN 1918	by **NIMBUS** 1910, a stayer, beat Bruleur over 12f in France; it is said that he was unlucky not to win the dreadful Derby of 1913 because he was hindered by the disaster at Tattenham Corner, but he was probably not the only one to have been upset in the fracas. Before being exported to Germany he sired †Le Capucin FD, Cerfeuil sire of the dam of the filly Samos AT. His sire was ***Elf II** 1893 GC and his dam **Nepthe** by Flying Fox (dam also of Nuage GP). Aileen won 11 flat and 4 hurdle races, imp. UK 1928 from France, bred 6 winners.
Ex YVELINE 1913 FO	by GARDEFEU p. 178.
Ex PHOTIME	by Delaunay.
Ex PRIMA VISTA	by St. Bris and 3 generations back to VIOLET – see Mrs. Butterwick p. 13. Family no. 8. Prima Vista was dam of Priori AT.

Schiaparelli won 5 small races, 12 and 13f.

She bred 11 foals, 8 of them won 24 races and £24,600.

SCHIAPARELLI 1935

Whose dam came from France

Mated with:–

Produce:–

KING SALMON

1940 HERRINGBONE 1, L,
Won 4 war-time races altogether, bred 8 wnrs. of 25½ races.

PIN STRIPE 1947
Hyperion
did not win, bred 4 wnrs.

LA FRESNES 1953
Court Martial
won 4 good races, bred 5 wnrs.

FOTHERINGAY
1964 Right Royal V
won once

RAGSTONE 1970
GC, Ragusa
Inbred Hyperion 4th rem.

FAIRWAY

1943 CUTAWAY
did not win,
bred 6 wnrs.

SCOLLATA
1952
Niccolo dell'Arca
won once

IRISH LASS II
1962
Sayajirao
exp. France 1962

IRISH BALL
1968 IRD,
Baldric II
exp. Japan 1974

HYPERION

1950 BARLEYCORN
Exp. France 1954
Did not win.

***†SHANTUNG**
1956 Sicambre
Won once over 7.5f as a 2yo and twice (10 & 12f) at 3 years. Definitely unlucky, rushing up to get 3rd place in Parthia's Derby, many observers being convinced that he would have won had he not been crossed and snatched up about a mile from home. Imported to UK 1960, he has sired Full Dress II 1, Genevra O and the dams of Nocturnal Spree 1 and Bruni L.

DAME D'ATOUR II
Cranach

RUTA 1960
Ratification

SASSAFRAS imp. UK '70
1967 FD, AT (3y) (FL)
Sheshoon

Part Two
THE SIRES

There is a touch of magic in the strings of fathers and sons which follow, but there is nothing in them to gainsay any known genetic truths, nor does the fact that they all descend from three sources have any bearing on their past excellence or future prospects.

BUT these are the stallions responsible for nearly all the classic races of the last 90 years and if those races in the years to come are won – which is unlikely – by horses connected in tail-male much further back towards the three founders, say *before* Teddy, Phalaris, Prince Palatine, Rabelais, Rialto, Blandford, Hyperion, Hurry On, Commando, Man o'War or Tourbillon, their succession will still remain of more than romantic interest and the study of whence they derive their virtue even more intriguing.

I have stated earlier on that remote tail-male ancestry has no bearing on current performance; as with the mares, it is the injection of outside blood over the generations that counts. By which I mean that, in the case of Cyllene (pp. 9, 151), the fact that he was line-bred to Birdcatcher did not make the slightest difference, whereas his inbreeding to Stockwell may well have done. Again, the more recent Nijinsky (pp. 156) must surely have been dependant for his great merit, not on his two lines of Bend Or, but on the vast racing prowess of his sire and dam.

So here follow, after a sketch of their tail-made origins (the Ancient Dynastics pp. 147), the surviving classic lines:– the dynasties of Bend Or 1877, St. Simon 1881, Isonomy 1875, Hampton 1872, West Australian 1850 and the French Gardefeu 1895.

THE ANCIENT DYNASTIES

Leading to today's surviving sire lines

THE DARLEY ARABIAN 170?
Childers 1715
Squirt 1732
Marske 1750-79
ECLIPSE 1764-89

- Pot-8-os 1773-1800
  Waxy 1790-1818
  Whalebone 1807-31
  - Camel 1822
    Touchstone 1831
    - Orlando 1841
      Eclipse (USA) 1849
      Led to
      Black Toney
      High Time , etc.
    - Newminster 1848
      HAMPTON 1872
      Led to
      Right Royal V
      St. Paddy
      Vaguely Noble
  - Sir Hercules 1826
    Birdcatcher 1833
    - The Baron 1842
      Stockwell 1849
      Doncaster 1870
      BEND OR 1877
      Led to
      Pharis II
      Tantieme
      Never Bend
      Hail to Reason
      Ballymoss
      Nijinsky
      Sea Bird II
      Queen's Hussar
    - Oxford 1857
      Sterling 1868
      ISONOMY 1875
      Led to
      Crepello
      Persian Gulf
      Vieux Manoir
- King Fergus 1775-1801
  Hambletonian 1792
  Whitelock 1803
  Blacklock 1814-31
  Voltaire 1826-48
  Voltigeur 1847-74
  Vedette 1854
  - Galopin 1872-99
    ST. SIMON 1881-1908
    - **PERSIMMON**
      1893
      Led to
      Princequillo
      Prince Chevalier
      Sicambre
    - **RABELAIS**
      1900
      Led to
      Ribot
      Worden
    - **CHAUCER**
      1900
      Led to
      Vatellor
      Bois Roussel
  - **Speculum** 1865
    Led to
    Phil Drake

THE BYERLEY TURK 168?
Jigg 170?
Partner 1718-47
Tartar 1743-59

HEROD 1758-79

Woodpecker 1773-98
Buzzard 1787-1811

- Castrel 1801-28
  Pantaloon 1824
  Windhound 1847
  Thormanby 1857
  Atlantic 1871
  Le Sancy 1884
  Le Samaritain 1895
  Roi Herode 1904
  The Tetrarch 1911

  Line faded out in
  classic racing
- the brothers
- Selim 1802-25
  Sultan 1816
  Bay Middleton 1833
  The Flying Dutchman 1846
  Dollar 1860
  Androcles 1870
  Cambyse 1884
  GARDEFEU 1895

  Led to
  Lorenzaccio
  Blakeney

THE GODOLPHIN ARABIAN 1724-53
Cade 1732-56

MATCHEM 1741-81

Conductor 1767
Trumpator 1782-1808
Sorcerer 1796
Comus 1809
Humphrey Clinker 1822-34
Melbourne 1834-59
West Australian 1850

- Solon 1861
  BARCALDINE 1878
  Led to
  Sheshoon
- Australian 1858 (exp. USA)
  Spendthrift 1876
  HASTINGS 1893
  Led to
  War Admiral
  Relic

THE DARLEY ARABIAN
to
ECLIPSE
to
BEND OR

— Bona Vista see later

1883 to 1904 *ORMONDE 2,D,L,Ch.

Unbeaten also in twelve other races, he ended up winning the 6f July Cup. He was sold to Argentina in 1889 for £12,000 and, in 1893, sold on to USA for 3 times that amount; there in 11 seasons he left only 16 foals. His UK offspring of note are Orme (below), Goldfinch (sire of Chelandry, p.39, 180, App. D), and Orville. Of his sparse USA get the only one to appear so far in European Classics is Ormondale who sired the 3rd dam of Dark Star and the 4th dam of Bald Eagle. Ormonde was a roarer, inherited, it is said, from Miss Agnes.

By *†BEND OR 1870 D, Ch.	*DONCASTER 1870 D, GC	Stockwell	The Baron to Eclipse
			Pocahontas
		Marigold	Teddington
			b mare 1852 Ratan
	ROUGE ROSE 1865	Thormanby	Windhound
			Alice Hawthorn
		Ellen Horne	Redshank
			Delhi
Ex LILY AGNES dam also of Ormonde's sister Ornament q.v.p.7	*†MACARONI 1860 2,D	Sweetmeat	Gladiator to Herod
			Lollypop
		Jocose	Pantaloon
			Banter
	POLLY AGNES	The Cure	Physician
			Morsel
		Miss Agnes	Birdcatcher
			Agnes by Clarion

1889 to 1915 *†ORME E (twice), Ch.

Won 14 races in all. He was poisoned before the 2000 which he probably would have won, nor could he run in the Derby which he probably would not have won as it seems he did not stay beyond 10f, see p.71, 148, App. D.

1904 to 1918 *†ORBY D, (Ir.D) see p.5, 148, App. D.

The most important of his get was the mare Orlass p.5, App. D.

By *ORMONDE 1883 2, D, L, Ch.	*†GALOPIN 1872 D	Vedette	Voltigeur to Eclipse
			Mrs. Ridgeway
		The Flying Duchess	The Flying Dutchman
			Merope
Ex ANGELICA 1879 St. Simon's sister; did not run, ancestress of Hill Gail & Superman (USA)	ST. ANGELA 1865	King Tom	Harkaway
			Pocahontas
		Adeline	Ion
			Little Fairy

1896 to 1911 *†FLYING FOX 2,D,L,E,

Won also 5 other races including the Jockey Club Stakes. He was sold to France in 1900 where he sired Ajax (his successor in the dynasty), 3 classic winners and 7 other good stallions; his daughters bred Nuage GP, Nimbus (1910) and *Radames

and now this line moves to France for six generations – over

By *†ORME 1889 E(twice), Ch. – – – – – GALOPIN			
Ex VAMPIRE 1889	*†GALOPIN 1872 D	Vedette	Voltigeur to Eclipse
			Mrs. Ridgeway
		The Flying Duchess	The Flying Dutchman
			Merope
	IRONY 1881 exp. USA '94	Rosebery	Speculum
			Ladylike
		Sarcasm	Breadalbane
			Jeu d'Esprit p. 95

Won twice at 2y, dangerously bad-tempered. Inbreeding to Galopin due to her being unsafe to send away to foal, so she was put to the home-sire Orme with Flying Fox as the result; this mating was repeated six times in all (with two changes to Grey Leg, another of the Duke of Westminster's home-sires) ending up with the production of *Vamose* in 1905 who was sent to France and became sire of the 4th dam of Royal Palace 2,D,E,GVI. Vampire had altogether 13 foals to live; her first she destroyed, then came *Batt* by Sheen, 2nd in the Derby and, after visiting S.America for ten years, returned and was sent to France in 1910. He appears in the hinterland of *Emerson's pedigree. Other daughters of Vampire are ancestresses of Pall Mall 2 and Aldford.

NB. DATES unless otherwise specified, are DATES OF FOALING.

BEND OR

Ormonde
Orme
Flying Fox

Bona Vista
see later

1901 *†AJAX FD,GP.

Unbeaten, won also Prix Lupin and other big races. The first classic winner he sired was Union 1906 FO (3rd dam of Le Pacha FD, GP, AT, Lied, Labrador and Tambourin). After that he got no classic winner until Teddy (next), though his son Chut acquired merit by siring the mare Quoi FO. Ajax's daughters bred *†Havresac II, *†Massine GC, AT, Invershin GC (twice) and Le Correge FD.

1913 *†TEDDY FD
(see p.129)

1923 *†ASTERUS Ch.(F2) see St. Astra p. 85.
If his renown has now faded in masculine circles, it still remains great through his daughters.

1922 *†AETHELSTAN

Won 9 races of up to 13f. He was said to be good-looking despite turned out toes. He sired *Maurepas GP (p.102), *Deiri (FL) (next) and *†Fantastic (FL), all stayers. He finished his days in USA.

By *† FLYING FOX 1896, 2, D,L,E. exp. France 1900 — GALOPIN
- GALOPIN
- †CLAMART 1888 GP sired d. of Finasseur FD,GP
  - Saumur — Dollar to Herod / Finlande FO
  - Princess Catherine — Prince Charlie / Catherine

Ex AMIE 1893 did not run, bred next year Adam (bro. to Ajax) exp. USA 1906 and went to Austria 1908. Sired Marsa FO
- Alice 1887
  - Wellingtonia — Chatanooga / Araucaria
  - Asta d. of Arreau GP,(F2) — Cambuslang / Lady Superior ex Penance, anc. of Timothy mare (see Double life p. 139).

By *†AJAX 1901 FD,GP by now recognised as a classic sire
- *BAY RONALD 1893 see p. 172.
  - Hampton — Lord Clifden to Eclipse / Lady Langden
  - Black Duchess — Galliard / Black Corrie

Ex RONDEAU 1900 ran 3 seasons winning well, exp. 1906 from UK to France where she lost 4 foals and bred two useless ones before having Teddy.
- Doremi 1894
  - Bend Or — Doncaster / Rouge Rose
  - Lady Emily — Macaroni / May Queen, anc. of Rosedrop O (Gainsborough's d.) and of Goyama and Dhaudevi GP.

By *†TEDDY 1913 FD — HAMPTON
- VAL SUZON 1905 sire of jumpers
  - Childwick — St. Simon to Eclipse / Plaisanterie
  - Thames Valley — HAMPTON / Rosy Brook

Ex DEDICACE 1915
- Disadvantage 1899
  - St. George — Cremorne / Strategy
  - Vantage — Hayden Edwards / Come and Go (American) and back 5 gens. to Galopade 1828 who was sent there.

BEND OR

Ormonde
Orme
Flying Fox
Ajax
Teddy
Aethelstan

1928 *DEIRI (FL)

A stayer who could not really be called a classic horse; however, he sired 3 classic winners: – *†Deux Pour Cent GP (next), Pirette FO (dam of Primera – p. 108), Bey FD and the useful sire *Majano.

By *†AETHELSTAN 1922			
Ex DESRA 1920 exp. from UK to France 1926	†CORCYRA 1911 see p. 97.	Polymelus	Cyllene to Bend Or
			Maid Marion
		Pearmain	Persimmon
			Nenemoosha
	Desna 1913 exp. Australia 1922	Desmond	St. Simon
			L'Abbesse de Jouarre O
		Dympna	St. David
			Court Card

1941 *†DEUX POUR CENT GP.

Another stayer. Exported to Czechoslovakia 1954. It cannot be said that he was disappointing as a sire since he was the father of the brilliant Tantieme, otherwise I have him on record only as the maternal grandsire of Tapalque FD.

By *DEIRI 1928 (FL)			
Ex DIX POUR CENT 1933 did not run	FERIDOON 1925 (exp. Fr.'28) unraced younger bro. to *†Diligence (sire of d. of Sun Chariot 1,0,L)	Hurry On	Marcovil to Matchem
			Toute Suite
		Ecurie	Radium
			Cheshire Cat
	La Chansonnerie 1923	Mesilim	Sans Souci II
			Malatesta
		La Francaise 4th d. of Tambourin II IRD and Nasram GVI	Simonian
			Keltoum ex Kendal Beauty exp. France 1898.

1947 to 1966 *†TANTIEME AT (twice), (F2)

Won also Lupin, Ganay, Coronation Cup etc. 12 races in all *and* he lost the FD by a short head. Up to 1950 he must have been the best horse in Europe since the war; he was blessed with a tremendous turn of speed and, descended as he was from a line of stayers, it is worth a glance at his pedigree to try and discover the origin of that most desirable aptitude. The only sire in sight not renowned (not always, I admit) for the manufacture of stayers up to Leger distance is his maternal grandsire the French 2000 winner, Indus (he sired also the 2nd dam of Bella Paola 1,0,Ch.). But Tantieme, with his fleetness of foot, did nothing to shorten the range of his descendants; he sired *Match III GVI, (FL & Wash. Int.), La Sega FO and her younger brother Danseur GP and, not without chances of carrying on this line, Match's younger brother Reliance FD,GP,(FL) and **Tanerko* (sire of Relko D and White Label GP). Tantieme's daughter Tantaliser bred Provoke L.

By *†DEUX POUR CENT 1941 GP			
Ex TERKA III 1942 did not win bred 7 wnrs.	†INDUS 1928 (F2)	Alcantara	Perth to Eclipse
			Toison d'Or
		Himalaya	Sardanapale
			Mountain Lass
	La Furka 1927 won 9 races	Blandford	Swynford
			Blanche
		Brenta	Sans Souci II
			Beaute de Neige d. of Mont Bernina, *Teleferique and 13 wnrs. back to Bella exp. France 1885

and now back to UK and the other Bend Or line – over

BEND OR

1889 — *†BONA VISTA 2

Stayed no further than a mile; besides Cyllene (next) he sired Lady Chancellor, dam of Challacombe L & 3rd dam of the American stallion Supremus. Bona Vista was exported to Austria 1897.

By BEND OR 1877 D,Ch — — — POCAHONTAS

Ex VISTA 1879 d. also of Sir Visto D,L & Velasquez E, Ch. (twice)

- *†MACARONI 1865 2, D the same process – a Macaroni mare mated with Bend Or – which produced Ornament p. 7 and Ormonde p. 148.
  - Sweetmeat: Gladiator to Herod / Lollypop
  - Jocose: Pantaloon / Banter
- Verdure 1867
  - King Tom: Harkaway / POCAHONTAS
  - May Bloom: Newminster / Lady Hawthorn 1954, bred 5 founder daughters

---

1895 — *†CYLLENE GC see Charm p. 9.

Exported to Argentina 1908

By *†BONA VISTA 1889 2 — — — STOCKWELL

Ex ARCADIA 1887 won two good 5f races at 2y.

- *†ISONOMY 1875 GC see p. 25.
  - Sterling: Oxford to Eclipse / Whisper
  - Isola Bella: STOCKWELL / Isoline
- DISTANT SHORE 1880 d. also of the bros. †St. Hilaire & St. Damien, also of Gulliver
  - Hermit: Newminster / Seclusion
  - Lands End: Trumpeter / Faraway ex Maid of Masham p. 11.

1902 to 1924 — *†POLYMELUS Ch. see p. 75.

By *†CYLLENE 1895 GC

Ex MAID MARION 1886 did not win, anc. also of Pont L'Eveque D, *†Wild Risk, *Right Royal V GVI & *Sea Hawk II.

- *†HAMPTON 1872 see p. 7.
  - Lord Clifden: Newminster to Eclipse / The Slave
  - Lady Langden: Kettledrum / Haricot
- QUIVER 1872 see p.49, 166, 168.
  - Toxophilite: Longbow / Legerdemain
  - b mare 1860: Y. Melbourne / Brown Bess 1844 founded 2 important families

---

1913 to 1931 — *†PHALARIS see p. 111.

Sickle & Fairway see later

By *†POLYMELUS 1902 Ch.

Ex BROMUS 1905 won once d. also of †Hainault p.29

- *†SAINFOIN 1887 D sired also *†Rocksand 2,D,L and the dams of *†Hurry On L and Franklin Ch. inbred to Stockwell at 3rd rem.
  - SPRINGFIELD: St. Albans to Eclipse / Viridis
  - Sanda: Wenlock / Sandal
- Cheery 1892 exp. Ger. 1913
  - St. Simon: Galopin / St. Angela
  - Sunrise g'dam of Eager and Grebe: SPRINGFIELD / Sunray ex Sunshine 1867 d. of Bushey Park & of 5 founder daughters

BEND OR

Bona Vista
Cyllene
Polymelus
Phalaris —— Sickle and Fairway (see later)

1920 to 1937 *†PHAROS Ch. see p. 56.

The first of a series of 9 classic winners by Phalaris from Chaucer mares. Whereas Phalaris sired 9 high class stallions, Pharos got eleven. Pharos was exp. Fr. '28

(& his bro. 1925 *†Fairway L,E,Ch. (twice)) see later

1936 to 1957 *†PHARIS II FD, GP, Unbeaten in 3 races. His dam was Carissima (granddam of Elpenor GC) by Clarissimus and he was inbred to Cyllene at the 4th remove. Champion sire in France twice, he was a P.O.W. of the Germans for the duration and was repatriated afterwards. He sired the winners of 8 classics and the important stallion *Pardal.

By *† PHALARIS 1913			ST. SIMON
		ST. SIMON	Galopin
			St. Angela
	*†CHAUCER 1900 p.38, (152), 156, 158, 167, 173, App. D.	Canterbury Pilgrim	Tristan
			Pilgrimage
Ex SCAPA FLOW 1914 p. 56.		Love Wisely	Wisdom
			Lovelorn
	ANCHORA 1904	Eryholme p.55-57, (152), 158, App. D.	Hazlehatch
			Ayrsmoss p. 55.

1935 to 1957 *†NEARCO GP see p. 100.

By *†PHAROS 1920 Ch.			ST. SIMON
		Rabelais	ST. SIMON
			Satirical
	†HAVRESAC II 1915 p. 99.	Hors Concours	Ajax
			Simona
Ex NOGARA 1928 p. 100.		Spearmint	Carbine
			Maid of the Mint
	CATNIP 1910 p. 99.	Sibola	The Sailor Prince
			Saluda

1940 to 1959 *†NASRULLAH Ch. p. 126.

of Nearco's first crop. exp. USA 1950

1951 to 1975 *NEVER SAY DIE D,L (see p.40) was the fifth American invader of European classics this century, the others having been: – 1901 Cap and Bells O, by a US sire from a UK mare exported to US 6 years earlier. 1907 Orby D (p. 5), English sire US mare imported UK 1896. 1912 Sweeper II 2 (p. 89) US sire, English Mare exp. US 1903. 1922 Pellsie FO, US sire imp. to France from a mare whose granddam was American. These were the vanguards; the main assault was yet to come and some of it can be seen in the next page or so.

By *†NEARCO 1935 GP		Blandford	Swynford
			Blanche
	*†BLENHEIM 1927 D p. 125.	Malva	Charles O'Malley
			Wild Arum
Ex MUMTAZ BEGUM 1932 p. 126.		The Tetrarch	Roi Herode
			Vahren
	MUMTAZ MAHAL 1921 p. 125.	Lady Josephine p. 103.	Sundridge
			Americus Girl p. 103.

BEND OR

Bona Vista
Cyllene
Polymelus
Phalaris —— Sickle & Fairway (see later)
Pharos
Nearco —— Royal Charger, Mossborough & Nearctic see later

1960 †NEVER BEND

Top of the US Free Handicap as a 2yo, then placed in both the Kentucky Derby and the Preakness Stakes; probably 10f was his limit. He has started his stud career well, being only 8 when MILL REEF was foaled. Never Bend's pedigree opposite appears to indicate just what his performances on the racecourse fulfilled, inherited speed from his sire and, from his dam's sire, speed again with about a 10f optimum. Mated with a Princequillo (stayer) mare the result was *MILL REEF 1968 D,GVI,E,AT, winner of 12 races out of 14. At least one book and goodness knows how many articles have been written about him and, for a really unbiased and erudite summary of his career, I recommend Timeform's "Racehorses of 1972". I saw nearly all his races including his 10 lengths victory in the Prix Ganay, but the one that impressed me most was his storming home by that distance in the York mud to win the Gimcrack as a 2yo. By now, standing at the National Stud, he has sired ACAMAS 1975 FD, (p. 90) SHIRLEY HEIGHTS 1975 D, IRD (p. 70).

- By *†NASRULLAH 1940 Ch.
- Ex LALUN 1952 won Kentucky O.
  - DJEDDAH 1945 E,Ch. p. 90.
    - Djebel
      - Tourbillon to Herod
      - Loika
    - Djezima
      - Asterus
      - Heldifann
  - Be Faithful 1942 won 14 races
    - Bimelech
      - Black Toney
      - La Troienne
    - Blood Root won 8 races
      - Blue Larkspur
      - Knockaney Bridge exp. USA 1924 p. 78.

1942 *ROYAL CHARGER

p. 126 exp. USA 1952

- By *†NEARCO 1935 GP
- Ex SUN PRINCESS 1937 p.
  - *†SOLARIO 1922 L,GC p. 106.
    - Gainsborough
      - Bayardo
      - Rosedrop
    - Sun Worship
      - Sundridge
      - Doctrine
  - MUMTAZ BEGUM 1932 v.s., dam of Nasrullah
    - Blenheim
      - Blandford
      - Malva
    - Mumtaz Mahal
      - The Tetrarch
      - Lady Josephine p. 103.

BEND OR

- Bona Vista
- Cyllene
- Polymelus
- Phalaris —— Sickle & Fairway see later
- Pharos
- Nearco —— Mossborough & Nearctic p.112, 155, App. D.
- Royal Charger

1951 to 1973 *TURN TO

A miler, retired hurt as a 3yo when shaping well. Imported to USA from France 1952

1959 *SIR GAYLORD see p.98, probably best at a mile. He is here because he sired the exceptional Sir Ivor 1965 2,D,Ch.(Wash.Int.), this he did from a mare by a stayer, *Mr. Trouble,* from a staying family going back to the Kentucky Oaks winner Alcibaiades 1927. He sired also the miler **Habitat* who has now entered the classic lists having sired Rose Bowl Ch. from the FO winner Roseliere. Sir Gaylord was exported to France from USA in 1973 and Habitat from USA to UK in 1967

1958 to 1976 *†HAIL TO REASON

Ran only as a 2yo, winning 9 races and coming out top of the USA Free Handicap. We meet him here as the sire of *Roberto* 1969 D and of the mare Priceless Gem (won 7 big races) the dam of Allez France FO,AT etc. He sired also winners of the Belmont and Preakness Stakes and the Kentucky Derby. Roberto, winner in all of 7 big races, comes from a Nashua (won 22 races) mare of a wonderful US line of winner-producers. He is at stud in the States. (exp. '73)

1947 to 1971 *†MOSSBOROUGH see p. 83, App. D.

By *ROYAL CHARGER 1942		PHAROS	
	*†ADMIRAL DRAKE 1931 GP p. 46.	Craig An Eran	Sunstar to Eclipse
			Maid of the Mist
		Plucky Liege	Spearmint
			Concertina
Ex SOURCE SUCREE 1940 won once at 2y, bred 9 winners, imp. UK, exp. USA '53	LAVENDULA II 1930 exp. France 1936, dam of †Ambiorix and 2nd d. of *†My Babu 2 and *Sayani.	PHAROS	Phalaris
			Scapa Flow
		Sweet Lavender 1923 exp. Fr. '28	Swynford
			Marchetta p. 133

			PHAROS
By *TURN TO 1951			PHAROS
			PLUCKY LIEGE
	BLUE SWORDS 1940 won 5 mid. dist. races	Blue Larkspur	Black Servant to Eclipse
			Blossom Time
		Flaming Swords	Man o'War
			Exalted
Ex NO THIRD CHANCE 1948	Gala Colors 1943	Sir Gallahad III	Teddy
			PLUCKY LIEGE
		Rouge Et Noir	St. Germans
			Baton Rouge

By *†NEARCO 1935 GP			CHAUCER
	BOBSLEIGH 1932 p.54, 153	Gainsborough	Bayardo to Eclipse
			Rosedrop
		Toboggan	Hurry On
			Glacier
Ex ALL MOONSHINE 1941 p. 83.	SELENE 1919 dam of *†Hyperion D,L, *Pharamond, *Sickle	CHAUCER	St. Simon
			Canterbury Pilgrim
		Serenissima	Minoru
			Gondolette p. 81

BEND OR

Bona Vista
Cyllene
Polymelus
Phalaris ——— Sickle & Fairway see later
Pharos
Nearco

1954 *†BALLYMOSS L,AT,E,GVI,IRD

Won 7 big races in all. His and his sire's hopes of survival as classic ancestors rest at present with Royal Palace 1964 2,D,E,GVI. (p. 20). Ballymoss sired also the dam of Levmoss GC, AT and his sister Sweet Mimosa FO.

By *†MOSSBOROUGH 1947 — — — GAINSBOROUGH

Ex INDIAN CALL 1936
did not win
bred 8 wnrs. of 28 races and £114,900.

*†SINGAPORE 1927 L
p. 14.
- GAINSBOROUGH
  - Bayardo
  - Rosedrop
- Tetrabazia
  - The Tetrarch
  - Abbazia

Flittermere 1926
- Buchan
  - Sunstar
  - Hamoaze
- Keysoe L
  - Swynford
  - Keystone 1903 O

A repeat of his sire's breeding pattern, by a Nearco horse, from a mare by a Gainsborough horse.

1954 to 1973 *NEARCTIC

Canadian miler, won 21 races. Northern Dancer (next) was of his first crop, thereafter he got countless stakes winners at home and in USA; then, in 1974 his influence returned to Europe in the form of Nonoalco 2.

By *†NEARCO 1935 GP — — — CHAUCER

Ex LADY ANGELA 1944
exp. Canada 1953

*†HYPERION 1930 D,L
- Gainsborough
  - Bayardo
  - Rosedrop
- Selene
  - CHAUCER
  - Serenissima

SISTER SARAH 1930
Bred 8 wnrs. of 30 races and £19,700
- Abbot's Trace
  - Tracery
  - Abbots Anne
- Sarita
  - Swynford
  - Molly Desmond p. 74.

Again the pattern repeated as in Mossborough and his son Ballymoss – by Nearco from a Gainsborough line mare.

1961 *NORTHERN DANCER

Foaled 27th May, the first foal of his dam. Won 14 races including the Kentucky Derby and the Preakness stakes. Sired also *The Minstrel* 1974 D, IRD, GVI

By *NEARCTIC 1954

Ex NATALMA 1957

*NATIVE DANCER 1950
won Preak., Bel. & 21 out of 22 races all told, sired Kauai King (Ken D & Preak.) & Hula Dancer 1, Ch.
- Polynesian
  - Unbreakable, grd. son of Phalaris
  - Black Polly
- Geisha
  - Discovery
  - Miyako

Almahmoud 1949
- Mahmoud
  - Blenheim
  - Mah Mahal
- Arbitrator
  - Peace Chance
  - Mother Goose and back to Princess Iskra exp. USA 1893

BEND OR

Bona Vista
Cyllene
Polymelus
Phalaris —— Fairway – see later

1967 *NIJINSKY 2,D,L,IRD,GVI(3yo).

For an account of him Timeform's "Racehorses of 1970" is recommended. imp. UK 1968, exp. USA 1971

By *NORTHERN DANCER 1961			
Ex FLAMING PAGE 1959 won Canadian Oaks granddam of The Minstrel 1974 D	†BULL PAGE 1947	Bull Lea	Bull Dog
			Rose Leaves
		Our Page	Blue Larkspur
			Occult by Dis Donc
	Flaring Top 1947	Menow	Pharamond
			Alcibaiades
		Flaming Top	Omaha
			Firetop and back to Torpenhow exp. USA 1915

The relative youth of the last two sires is perhaps worth noting.

1924 *SICKLE

Exported from UK to USA 1929 after a not very brilliant racing career; he was probably best described as a miler. In the States he was champion sire twice. Within the limits of this book he is met again as the sire of the dam of Dan Cupid, the latter being formidably inbred, having Sickle at the 2nd and 4th removes (see later).

By *†PHALARIS 1913	*†POLYMELUS 1902 Ch.	CYLLENE	Bona Vista
			Arcadia
		Maid Marion	Hampton
			Quiver
	BROMUS	Sainfoin	Springfield
			Sanda
		Cheery	ST. SIMON
			Sunrise
Ex SELENE 1919 d. also of *Pharamond (bro. to Sickle), Hunter's Moon and Moonlight Run and *†Hyperion D,L	*†CHAUCER 1900	ST. SIMON	Galopin
			St. Angela
		Canterbury Pilgrim	Tristan
			Pilgrimage
	SERENISSIMA 1913	Minoru	CYLLENE
			Mother Siegel
		Gondolette p. 81	Loved One
			Dongola

BEND OR

Bona Vista
Cyllene
Polymelus
Phalaris —— Fairway – see later
Sickle

1935 UNBREAKABLE

Won 5 good races in England in 3 seasons, his best distance was 7f to a mile. Successful sire in USA

By *SICKLE 1924			SAINFOIN
			ST. SIMON
Ex BLUE GRASS 1917	*PRINCE PALATINE 1908 L,GC (twice) E p.89	Persimmon	ST. SIMON
			Perdita II
		Lady Lightfoot	Isinglass
			Glare
	Hour Glass II 1909	Rock sand	SAINFOIN
			Roquebrune
		Hautesse exp. USA 1903	Archiduc
			Hauteur 1, exp. France 1884, bred foals there and ret. UK 1890

1942 *POLYNESIAN

Won the Preakness Stakes and other great races in 4 seasons, he ran 59 times in all and became champion sire in USA. He sired also Polly's Jet (won 9 big, short races) and the dam of Right Tack 2 (Ir.2).

By UNBREAKABLE 1935			POLYMELUS
Ex BLACK POLLY 1936	POLYMELIAN 1914 Exp. USA '14	POLYMELUS	Cyllene
			Maid Marion
		Pasquita 1907 exp USA '19	Sundridge
			Pasquil
	Black Queen 1930	Pompey	Sunbriar
			Cleopatra
		Black Marie	Black Toney
			Bird Loose 1916 probably exp. from France to USA

1950 to 1957 *NATIVE DANCER

Won Preakness, Belmont and 21 out of 22 races all told. See p. 155 (Northern Dancer)

By *POLYNESIAN 1942			
Ex GEISHA 1943 won once bred 4 wnrs.	†DISCOVERY 1931 mid. dist. won 27 races in 4 seasons	Display	Fair Play to Matchem
			Cicuta
		Ariadne	Light Brigade
			Adrienne
	Myako 1935	John P. Grier	Whisk Broom II
			Wonder
		La Chica	Sweep
			La Grisette exp to USA 1925

BEND OR

Bona Vista
Cyllene
Polymelus
Phalaris

1956 to 1972 — *DAN CUPID

2nd by a short head in Herbager's FD. His average winning distance was under a mile.

- By *NATIVE DANCER 1950 — SICKLE
- Ex VIXINETTE 1944
  - *SICKLE 1924
    - Phalaris
      - Polymelus
      - Bromus
    - Selene
      - Chaucer
      - Serenissima
  - Lady Reynard 1939
    - Gallant Fox
      - Sir Gallahad III
      - Marguerite
    - Nerva
      - Fair Play
      - Zephyretta whose gt.grd. dam was the US matriarch Sallie McLelland 1888

1962 to 1973 — *SEA BIRD II D,AT, (Lupin etc.) see p.6

Unfortunately died aged only 11; he has some good sons in USA, also there is Gyr.

- By *DAN CUPID 1956
- Ex SICALADE 1956 did not win
  - *†SICAMBRE 1948 FD, GP p. 96
    - Prince Bio
      - Prince Rose to Eclipse
      - Biologie
    - Sif
      - Rialto
      - Suavita
  - Marmalade 1949 did not run
    - Maurepas
      - Aethelstan
      - Broceliande
    - Couleur d. of Camaree 1
      - Biribi
      - Colour Bar exp . France 1936

1925 to 1948 — *†FAIRWAY L,E,Ch.(twice) see p. 56

Brother to Pharos

- By *†PHALARIS 1913
  - *†POLYMELUS 1902 Ch.
    - Cyllene
      - Bona Vista
      - Arcadia
    - Maid Marion
      - Hampton
      - Quiver
  - BROMUS
    - Sainfoin
      - Springfield
      - Sanda
    - Cheery
      - ST. SIMON
      - Sunrise
- Ex SCAPA FLOW 1914
  - *†CHAUCER 1900
    - ST. SIMON
      - Galopin
      - St. Angela
    - Canterbury Pilgrim
      - Tristan
      - Pilgrimage
  - Anchora 1904
    - Love Wisely
      - Wisdom
      - Lovelorn
    - Eryholme p. 55
      - Hazlehatch
      - Ayrsmoss p. 55

BEND OR

Bona Vista
Cyllene
Polymelus
Phalaris
Fairway

Years	Horse	Parents	Grandparents	Great-grandparents	Great-great-grandparents
1932 to 1958	*†FAIR TRIAL see p. 104	By *†FAIRWAY 1925 L,E,Ch.(twice)			
		Ex LADY JUROR 1919	*†SON IN LAW 1911 p. 103	Dark Ronald	Bay Ronald to Eclipse
					Darkie
				Mother In Law	Matchmaker
					Be Cannie
			LADY JOSEPHINE	Sundridge	Amphion
					Sierra
				Americus Girl	Americus
					Palotta p. 103
1944 to 1964	*†PETITION E see p. 67	By *†FAIR TRIAL 1932			
		Ex ART PAPER 1933 won twice	ARTIST'S PROOF 1926 p. 94	Gainsborough	Bayardo
					Rosedrop
				Clear Evidence	Tracery
					Honora
			Quire 1918	Fairy King	Desmond
					Queen Fairy
				Queen Carbine	Carbine
					Sceptre p.8, 65-67, App. D.
1950 to 1976	MARCH PAST see p. 60	By *†PETITION 1944 E			
		Ex MARCELETTE 1942 did not win	WILLIAM OF VALENCE 1932 a good 10f handicapper, ran 17 times and won 5 races inc. the City & Suburban & Rosebery handicaps.	Vatout	Prince Chimay to Eclipse
					Vasthi
				Queen Iseult	Teddy
					Sweet Agnes
			Permavon 1936	Stratford	Swynford
					Lesbia
				Curl Paper	Papyrus
					Coila

And now his son PETINGO, from an Alycidon mare, has sired the 1974 IRD winner, English Prince and the 1979 D IRD and GVI winner Troy

BEND OR

Bona Vista
Cyllene
Polymelus
Phalaris
Fairway
Fair Trial
Petition
March Past

1960 *QUEEN'S HUSSAR

A very good miler indeed. He ran 21 times in all; at 2yrs. he won 4 times, at 3 he took the Lockinge Stakes at Newbury and the Sussex Stakes at Goodwood, at 4 he won one other 2 horse race.
As can be seen, he is very closely in-bred to Fair Trial and his father Fairway, enough Phalaris blood one would think, to frighten off many mares containing that element. Note too the speed factor introduced by the sprinter Vilmorin with his two close lines of Orby; like March Past, Vilmorin had never before appeared in a classic pedigree.
Before coming to the Brigadier who won 9 great races, I must point out that up to the moment of writing Queen's Hussar has sired the winners of 146 races worth over half a million £s, included among them being the excellent Shiny Tenth and the classic filly Highclere, winner of the 1000 gns. and the French Oaks.

By MARCH PAST 1950 — FAIR TRIAL

FAIRWAY

Ex JOJO 1950 won one 5f race at 3y; bred the wnrs. of over 40 races, the most notable being Queen's Hussar's gelding brother, Scots Fusilier who raced for 8 seasons, ran 74 times and won 22 races and £12,000 odd. His max. winning distance was 10f.

VILMORIN 1943 won King's Stand Stakes at 3y; sired mostly fast horses, the best being *Quorum**, Vilmoray and Vigo.
*Sire of Red Rum triple wnr. of the Grand National

Gold Bridge
- Golden Boss to Orby
- Flying Diadem to Orby

Queen of the Meadows
- FAIRWAY
- Queen of The Blues

Fairy Jane 1939

FAIR TRIAL
- FAIRWAY
- Lady Juror

Light Tackle
- Salmon Trout
- True Joy

I show opposite the pedigree of Brigadier Gerard 1968 2,E,GVI, Ch. (twice) if only for the reason that I inferred above that it would be a brave breeder who mated a mare containing Phalaris blood with Queen's Hussar.
If there could have been evil in this, one presumes that the 4th remove is sufficiently remote now that the other 3 strains have receded into the 6th generation. Anyway it worked.
Brigadier Gerard, now of course at stud, is big and very beautiful. I shall say no more about him save to mention that he was beaten only once, over his best distance of 10f, and that it is my opinion that that occasion at York came too soon after his 12f victory as a 4yo in the GVI.

By *QUEEN'S HUSSAR 1960 — FAIR TRIAL

FAIR TRIAL

Ex LA PAIVA 1956 did not win

*†PRINCE CHEVALIER 1943 FD p.

Prince Rose
- Rose Prince to Eclipse
- Indolence

Chevalerie
- Abbots Speed
- Kassala

Brazen Molly 1940

Horus
- Papyrus
- Lady Peregrine

Molly Adare
- Phalaris
- Molly Desmond p. 74

THE DARLEY ARABIAN

to

Eclipse

to

King Fergus

to

Blacklock

to

ST. SIMON

— Rabelais & Chaucer see later

1893 to 1908 *†PERSIMMON D,L,GC,E see p. 7

and his brothers: *†FLORIZEL II 1891 p.29

and DIAMOND JUBILEE 1897 2,D,L,E exp. Argentina 1906, his name appears in the background only of 2 classic winners.

By *†ST. SIMON 1881 GC unbeaten	*†GALOPIN 1872 D	Vedette	Voltigeur
			Mrs. Ridgeway
		The Flying Duchess	The Flying Dutchman
			Merope
	St. Angela 1865 exp. Fr. '83	King Tom	Harkaway
			Pocahontas
		Adeline	Ion
			Little Fairy
Ex PERDITA II 1881 won 7 races See Introduction	*†HAMPTON 1872 p. 7	Lord Clifden	Newminter to Eclipse
			The Slave
		Lady Langden	Kettledrum
			Haricot
	Hermione 1875	Y. Melbourne	Melbourne
			Clarissa
		La Belle Helene whose 5 daus. founded branches of family no. 7	St. Albans
			Teterrima

1908 to 1924 *PRINCE PALATINE L,GC(twice)E see p. 89

By *†PERSIMMON 1893 D,L,GC,E — — — — — — — — — — HAMPTON			
Ex LADY LIGHTFOOT 1900, won 6 mod. races, bred wnrs. of 20 races, ancestress of Random Shot GC.	*†ISINGLASS 1890 2,D,L,GC,p.13	Isonomy	Sterling to Eclipse
			Isola Bella
		Deadlock	Wenlock
			Malpractice
	GLARE 1891 dam of Flair 1	Ayrshire	HAMPTON
			Atalanta
		Footlight	Cremorne
			Paraffin

1919 *ROSE PRINCE

A long distance stayer who won the Cesarewitch, Queen Alexandra Stakes and 3 other races. He was bred in France from a line of first class mares, imported into UK in 1924 and sent back to France in 1930. Not a great success at stud, he got Prince Rose (next) who kept his and his sire's place in the lineal succession and †Mousson, a good winner in France and sire of Aglae Glace FO.

and now to France –

By *†PRINCE PALATINE 1908 L,GC(twice),E			
Ex EGLANTINE 1906	*†PERTH 1896 FD,GP, (F2,FL) 3 times champion sire in France, sired among other notabilities Alcantara II FD, Northeast GP & the d. of *†Belfonds FD	War Dance	Galliard to Eclipse
			Warpaint
		Primrose Dame 1885 exp. France 1894	Barcaldine
			Lady Rosebery
	ROSE DE MAI 1900 FO,(F1) 4th dam of Aglae Glace FO	Callistrate	Cambyse
			Citronelle
		Maypole (F1)	Silvio
			Merry May 1873 exp. France 1879

ST. SIMON

Rabelais & Chaucer
see later

Persimmon
Prince Palatine
Rose Prince

1928 *PRINCE ROSE

Was bred in England and sold to Belgium as a foal; there and in France he ran 19 times in three seasons over middle distances and was beaten only 3 times. He stood in Belgium where unfortunately the quality of the mares were not worthy of him, until 1938 when he went to France to become champion through 3 sons, perhaps as influential in the history of racing as any three one can name: – Prince Bio (next), Prince Chevalier and Princequillo (next page).

- By *ROSE PRINCE 1919
- Ex INDOLENCE 1920 Won once bred 2 other winners.
  - *†GAY CRUSADER 1914 2,D,L,GC,Ch., all war-time wins at Newmarket; but he was one of the best horses ever. Inbred to Galopin at the 3rd remove. Not a success at stud except for his daughters which bred also *†Djebel 2,AT, †El Greco and Hierocles.
    - Bayardo
      - Bay Ronald to Eclipse
      - Galicia
    - Gay Laura
      - Beppo
      - Galleotia 1
  - BARRIER 1910 anc. of Rheingold AT, Barrier bred 5 wnrs.
    - Grey Leg
      - Pepper and Salt
      - Quetta
    - Bar The Way
      - Rightaway
      - Barrisdale

1941 *†PRINCE BIO (F2)

and 6 other races in 2 seasons in France; there he became champion sire, responsible for Sicambre (next), Northern Light GP, Le Petit Prince FD, Rose Royale II 1, Ch. and Prince Taj. His daughter Senones, Sicambre's sister, bred *Snob, sire of Goodly FD.

- By *PRINCE ROSE 1928
- Ex BIOLOGIE 1935 did not win
  - *BACTERIOPHAGE 1929, ran only once and died after 2 years at stud. Sired *Teleferique sire of *†Alizier sire of Tapalque FD on whom rests the survival of The Tetrarch's classic line.
    - Tetratema
      - The Tetrarch to Herod
      - Scotch Gift
    - Pharmacie 1918 exp. Fr. '28
      - Charles O'malley
      - Prescription
  - Eponge 1929
    - Cadum
      - Sans Souci II
      - Spring Cleaning
    - Sea Moss 1917 exp. France 1921 Grd.dam of Peniche FO
      - William the Third
      - Seadune

1948 *†SICAMBRE FD,GP
see p. 96

There seems just a hope that *†Shantung 1956 and *Diatome 1962 might carry on this line.

- By *†PRINCE BIO 1941 (F2)
- Ex SIF 1936 won 3 good races
  - *†RIALTO 1928 p. 95
    - RABELAIS
      - St. Simon
      - Satirical
    - La Grelee
      - Helicon
      - Grignouse
  - Suavita 1928
    - Alcantara II
      - Perth
      - Toison d'or
    - Shocking
      - RABELAIS
      - Saperlipopette p. 95

ST. SIMON

— Rabelais & Chaucer see later

Persimmon
Prince Palatine
Rose Prince
Prince Rose

1943 to 1961 — *†PRINCE CHEVALIER FD see p. 86, 150, App. C, App. D.

By *PRINCE ROSE 1928			
	†ABBOTS SPEED 1923 mid. dist. h'cpr. won the Jubilee twice exp. France '28.	Abbots Trace	Tracery to Eclipse
			Abbots Anne
		Mary Gaunt	John O'Gaunt
			Quick
Ex CHEVALERIE 1923 won 2 flat and 2 hurdles	KASSALA 1926	Cylgad	Cyllene
			Gadfly
		Farizade 4th d. of Armistice GP	Sardanapale
			Diavolezza ex St. Astra FO p. 85

1940 *†PRINCEQUILLO, exp. USA '41, won the Preakness Stakes and 11 other big races. Sired Tambourine II IRD, *Round Table (won Ken.D and sired Baldric II 2, Ch. and the big American winner Hill Prince); his daughters have bred *Sir Gaylord, Mill Reef D,GVI,E, AT and San San AT.

By *PRINCE ROSE 1928			
	*†PAPYRUS 1920 D won 9 races all told, 2nd in Leger, sired Horus and the dams also of *†Honeyway Ch., Traghetto (It.D) & Northern Light GP.	Tracery	Rockstand to Eclipse
			Topiary
		Miss Matty, did not run, bred 9 wnrs.	Marcovil
			Simonath
Ex COSQUILLA 1933 did not win, exp. France 1935 and back to Ireland then to USA 1946	QUICK THOUGHT 1918 Bred 6 wnrs. of 25½ races	White Eagle	Gallinule
			Merry Gal
		Mindful	Minoru
			Noble Martha

ST. SIMON

— Chaucer – see later

1900 *†RABELAIS see p. 24

exp. France 1904

— Rialto – see later

1915 †HAVRESAC II p.99

1926 *CAVALIERE D'ARPINO

Unbeaten in 6 big Italian races. Sired Bellini (next) and Traghetto (It.D)

By *†ST. SIMON GC 1881	†SATIETY 1885 p.23	Isonomy	Sterling to Eclipse
			Isola Bella
		Wifey	Cremorne
			Lady Mary
Ex SATIRICAL 1891 p. 23	Chaff 1880	Wild Oats	Wild Dayrell
			The Golden Horn
		Celerrima	Stockwell
			Slander

By *†RABELAIS 1900	ST. SIMON		
	*†AJAX 1901 FD, GP p. 149	Flying Fox	Orme to Eclipse
			Vampire
		Aimie	Clamart
			Alice
Ex HORS CONCOURS 1906 did not win	Simona 1893	ST. SIMON	Galopin
			St. Angela
		Flying Footstep	Atalanta by Galopin, dam of Ayrshire 2,D,E & Grd. – dau. of Woodbine p. 13

By †HAVRESAC II 1915		ST. SIMON	GALOPIN
			ST. SIMON
	*†CICERO 1902 D p. 65	Cyllene	Bona Vista to Eclipse
			Arcadia
		Gas	Ayrshire
			Illuminata
Ex CHUETTE 1916 ran 22 times, won 7 exp. Italy 1920 d. also of Cranach 1923.	Chute 1905	Carbine	Musket
			Mersey
		Weir, anc. of Pourparler 1 & Fleet 1.	GALOPIN
			Dee, anc. of Diolite 2 & Lavandin D.

This line goes back to Otheothea 1772 ancestress of the mare Kincsem 1874 of undying memory, the Hungarian racing wonder whose line has been thoughtfully kept alive and, 100 years after her birth, produced the Oaks winner, Polygamy.

ST. SIMON

Chaucer – see later

Rabelais — Rialto - see later
Havresac II
Cavaliero d'Arpino

1937 BELLINI

(It.D,L), also won in Germany. At stud in Italy but captured by the Germans in 1944 and soon lost to the Russians.

By Cavaliere d'Arpino 1926			
BELLA MINNA 1923 exp. It. '35 p. 37, won 7 races	*†BACHELOR'S DOUBLE 1906 p. 79	Tredennis	Kendal
			St. Marguerite I
		Lady Bawn	Le Noir
			Milady
	SANTA MINA 1913	Santoi	Queen's Birthday
			Merry Wife
		Minnow	William The Third
			Mino

1944 *TENERANI

Won 15 races in Italy including their D & L, the Queen Elizabeth Stakes (12f) at Ascot & the Goodwood Cup. Went to stud in Italy first and then, in 1951, he was imported to the UK National Stud, but not before he had sired Ribot (next). In England he sired Fighting Charlie GC. He returned to Italy '60.

By BELLINI 1937			
Ex TOFANELLA 1931 sold as a yrlg. for 140 gns. & exp. Italy 1932. Won 4 good races, d. also of Tokamura (It,1,L) d. of Toulouse Lautrec & the wnrs. of 4 Italian classics	*†APELLE 1923 p, 131	Sardanapale	Prestige to Eclipse
			Gemma
		Angelina	St. Frusquin
			Seraphine
	Try Try Again 1922	Cylgad	Cyllene
			Gadfly
		Perseverance II	Persimmon
			Reminiscence

1952 to 1972 *RIBOT GVI,AT

(twice) & 14 other of the major races in Italy; he was never beaten. Imported to England in 1956 he had two seasons at stud; among his first crop was Molvedo 1958 AT (at stud in Italy), next year he got the good miler Romulus; then he returned to Italy. Among his 3rd crop was *Ragusa 1960 IRD,GVI,L,E, this made Ribot champion sire in UK. In 1960 he was sent to USA and there he got the English Oaks winner Long Look, to be followed by Prince Royal AT and the brothers Ribocco & Ribero, each winning both the Irish Derby and the Leger. He sired also Yrrah Jr (sire of Hurry Harriet Ch.) and *Graustark (sire of Caracolero FD); below comes Ragusa as a possible linear successor, but with no confidence, due to his death in 1973.

By *TENERANI 1944			
Ex ROMANELLA 1943 won 5 races at 2y bred 10 winners.	†EL GRECO 1934 won 17 mid. dist. races in Italy, sent to France 1949, died 1950	Pharos	Phalaris
			Scapa Flow
		Gay Gamp 1923 vis. It. '33, '35	Gay Crusader
			Parasol
	BARBARA BURRINI 1937, exp. Italy 1937 won 6 races	Papyrus	Tracery
			Miss Matty
		Bucolic	Buchan
			Volcanic

1960 to 1973 *RAGUSA IRD,GVI,L,E,

Went to stud after the 1964 season, sired Morston D and Ragstone GC.

By *RIBOT 1952 GVI, AT (twice)			PHAROS
Ex FANTAN II 1952 Imp. UK '56. exp. USA '63	†AMBIORIX 1946 Won Lupin, 2nd FD, sent to USA, died 1975	Tourbillon	Ksar to Herod
			Durban
		Lavendula II	PHAROS
			Sweet Lavender
	Red Eye 1936	Petee Wrack	Wrack
			Marguerite
		Charred Keg	Stimulus
			Jug of Wine

An American line which had not been concerned with European classics for over 100 years; it goes back to Maria Black, exported to USA, probably as a yearling, in 1835.

ST. SIMON

Chaucer – see later

Rabelais

1923 *†RIALTO see p. 95

By *†RABELAIS 1900

Ex LA GRELEE 1918 did not win 3rd d. of Tahiti FO

- †HELICON 1908 p.137 exp. France 1912
  - Cyllene
    - Bona Vista to Eclipse
    - Arcadia
  - Vain Duchess
    - ISINGLASS
    - Sweet Duchess
- GRIGNOUSE 1910 d. of Grand Guignol
  - Kilglass
    - ISINGLASS
    - Kilkenny
  - Simper
    - Sempronius
    - Value (USA)

I cannot trace which mare, possibly in the 18th century, introduced this line to America; from it stem Lexington, Commando, Stimulus & 3 European classic winners.

1940 *†WILD RISK

Won 4 out of 22 races in 4 seasons (12 to 20f) on the flat, but achieved much more over hurdles by winning the championship twice. A tough customer with an off-fore club-foot. He went to stud in France in 1946 and sired Vimy GVI, Worden (next), Balto GP, GC and Le Fabuleux FD.

By *†RIALTO 1923

Ex WILD VIOLET 1935 won 3 races

- *†BLANDFORD 1919 p. 36
  - Swynford
    - John o'Gaunt to Eclipse
    - Canterbury Pilgrim
  - Blanche
    - White Eagle
    - Black Cherry
- Wood Violet 1928 Imp. UK exp. Fr. '34
  - Ksar
    - Bruleur
    - Kizil Kourgan
  - Pervencheres 3rd d. of *Right Royal V FD, GVI
    - Maboul II
    - Poet's Star, gt.grd.dau. of Quiver p. 49 and gr. dam of Pont L'Eveque D.

1949 to 1969 *†WORDEN

Foaled Fr, imp. UK '59. Won the Washington International and 3 big European races, he suffered a desperate series of 2nds and 3rds including 2 thirds in the GVI. He has sired Barquette FO, Armistice GP and that very useful horse Bon Mot III AT (already the sire of Lassalle GC) who was unfortunately sent to Japan in 1972. Worden's daughters have bred so far Dhaudevi GP, Grundy D, IRD, GVI, (Ir.2), Juliette Marny O and Wollow 2,E, Julio Mariner L

By *†WILD RISK 1940

Ex SANS TARES 1939 did not run 3rd dam of Le Fabuleux FD

- †SIND 1933 p. 33
  - Solario
    - Gainsborough to Eclipse
    - Sun Worship
  - Mirawala
    - Phalaris
    - Miranda
- TARA 1932 dam of *†Norseman
  - Teddy
    - Ajax
    - Rondeau
  - Jean Gow 1921 exp. USA 1923
    - Neil Gow
    - Jane Shore

ST. SIMON

1900 to 1926 — *†CHAUCER see p. 38

As will be seen below his branch of St. Simon's dynasty has not survived beyond Tulyar 1949 and the chances of revival seem very remote. I include Chaucer, however, because of his enormous influence as a sire of brood-mares; I repeat the details:– he sired the dams of *†Pharos Ch., *†Fairway L,E,Ch.(twice), Fair Isle 1, Warden of the Marches Ch., *†Hyperion D,L, *Mossborough, Colorado 2,E, *†Caerleon E, *Colombo 2, *†Apron and *Sickle. His only classic winning son was *†Stedfast Ch. who sired the Oaks winner Brownhylda and also the dam of *Firdaussi and the 4th dam of *Queen's Hussar. Chaucer's other son Prince Chimay (next) provided a very weak link in the chain.

By *†ST. SIMON			
Ex CANTERBURY PILGRIM 1893 O p.37, 168, App. D.	*†TRISTAN 1878 GC,Ch.(three times) p. 11	Hermit	Newminster to Eclipse
			Seclusion
		Thrift	Stockwell
			Braxey
	PILGRIMAGE 1875 2, 1,	The Palmer	Beadsman
			Madame Eglantine
		Lady Audley	Macaroni
			Secret

1915 — PRINCE CHIMAY

Exported to France in 1919, won the Jockey Club Stakes (beating Gainsborough) and 3 other races. Bred in England he went to France after the war where he sired only one useful horse:–

By *†CHAUCER 1900			HERMIT
Ex GALLORETTE 1907 won 1 small race	*†GALLINULE 1884 p. 13	Isonomy	Sterling
			Isola Bella
		Moorhen	HERMIT
			Sister to Ryshworth
	Orlet 1891 3rd d. of Invershin GC (twice)	Bend Or	Doncaster
			Rouge Rose
		Ruth dam of Tarporley	The Scottish Chief
			Hilda

1926 to 1937 — *VATOUT (F2)

and 4 other races, went to stud aged 5 and died when only 11. Unlike his sire and grandsire his offspring included 4 very good males:– William of Valence (p.159), *†Bois Roussel D, *†Vatellor and Atout Maitre. His best daughter was Boreale, granddam of Never Say Die; another was 4th dam of Topyo AT. All of which looks encouraging for the continuance of the line; but the only ones in the running were Vatellor & Bois Roussel and this is what happened to their sons:– <u>Vatellor's</u> – Pearl Diver D went to Japan, My Love D,GP (a very exhausting double) achieved nothing at stud in France before being sold to Argentina, Vattel GP I cannot trace as having sired anything of note. <u>Bois Roussel's</u> – Migoli E,Ch.,AT was disappointing at stud, so was Ridge Wood L, leaving *Tehran L with the chance of carrying on. <u>Tehran</u> had 3 possible successors, Tabriz 1947 (sire of Taboun 2), *†Shikampur 1950 (sire only of the dam of Niksar 2) then the unfortunate Tulyar D,L,GVI– see p. 16 – who seems to have but small chances of reviving the Chaucer line.

By PRINCE CHIMAY 1915		GALLINULE	
Ex VASTHI 1921	†SANS SOUCI II 1904 p. 101	Le Roi Soleil	Heaume to Eclipse
			M'sllc de la Valliere
		Sanctimony 1896 Exp. France 1902	St. Serf
			Golden Iris
	VAYA 1909 Exp. France 1912 2nd d. of *†Vandale & †Verso II FD,AT,(FL)	Beppo	Marco
			Pitti
		Waterhen 3rd d. of Flowershop FO, (F1)	GALLINULE
			Gipsy Queen grd.dam of Quelle Est Belle FO

THE DARLEY ARABIAN
to
Eclipse
to
Pot.8.os
to
Whalebone
to
Sir Hercules
to
Birdcatcher
to
ISONOMY

1896 to 1911 *†ISINGLASS 2,D,L,GC,E see p. 13

With the exception of perhaps the worst horse he sired, which (see below) carried on the Isonomy stirp for him, Isinglass did more for racing through his mares than through his sons; of these the two which I have dealt with, Goody Two Shoes and Lady Lightfoot were outstanding.

1901 to 1924 *†JOHN O'GAUNT see p. 37

In addition to Swynford (next) he sired Kennymore 2, Harry of Hereford and the dams of Abbot's Speed & the American horse Supremus 1922.

1907 to 1928 *†SWYNFORD L,E see p. 36

Parents	Grandparents	Great-grandparents	Great-great-grandparents
By *†ISONOMY 1875 GC (twice) p. 25	*†STERLING 1868 p. 19	Oxford	Birdcatcher
			Honey Dear
		Whisper	Flatcatcher
			Silence
	ISOLA BELLA 1868	Stockwell	The Baron
			Pocahontas
		Isoline	Ethelbert
			Bassishaw
Ex DEADLOCK 1878 did not run; she was thrown in gratis when a previous owner bought a hunter.	*†WENLOCK 1869 L p. 20	Lord Clifden	Newminster to Eclipse
			Lady Langden
		Mineral	Rataplan
			Manganese
	Malpractice 1864	Chevalier d'Industrie	Orlando
			Industry O
		Dutchman's Daughter	The Flying Dutchman
			Red Rose

Parents	Grandparents	Great-grandparents	Great-great-grandparents
By *†ISINGLASS 2,D,L, GC,E, 1896	*†ST. SIMON 1881 GC	Galopin	Vedette to Eclipse
			The Flying Duchess
		St. Angela	King Tom
			Adeline
Ex LA FLECHE 1889 1,O,L,GC granddam of Cinna 1.	QUIVER 1872 p. 49	Toxophilite	Longbow
			Legerdemain
		b mare 1860	Y.Melbourne
			Brown Bess

Parents	Grandparents	Great-grandparents	Great-great-grandparents
By *†JOHN O'GAUNT 1901	*†TRISTAN 1878 GC,Ch. (3 times) p. 11	Hermit	Newminster
			Seclusion
		Thrift	Stockwell
			Braxey
Ex CANTERBURY PILGRIM 1893 O p. 37	PILGRIMAGE 1875 2, 1.	The Palmer	Beadsman
			Madam Eglantine
		Lady Audley d. of St. Louis and Pellerino	Macaroni
			Secret

ISONOMY

Isinglass
John O'Gaunt
Swynford

1919 to 1935 *†BLANDFORD see p. 36

Brantome & Bahram see later

1928 to 1958 *†BLENHEIM D see p. 125

Went to stud in France 1932 and USA 1936

1934 to 1955 *†DONATELLO II (It.D) see p.74, 170, 173, App. C, App. D.

He sired Picture Play 1 (3rd dam of Royal Palace 2,D,E,GVI), Alycidon (next), Crepello (next page), Supertello GC, †Acropolis (Alycidon's brother) and the dams of *Pinza D,GVI, *AUREOLE GVI and *High Hat.

1945 to 1963 *†ALYCIDON GC see p. 135

By *†SWYNFORD 1907 L,E — ISONOMY

Ex BLANCHE 1912 did not win
- †WHITE EAGLE 1905 p. 35
  - Gallinule: ISONOMY, Moorhen
  - Merry Gal: Galopin, Mary Seaton
- BLACK CHERRY 1892 p. 35
  - Bendigo: Ben Battle, Hasty Girl
  - Black Duchess d. of *Bay Ronald: Galliard, Black Corrie

By *†BLANDFORD 1919 — ISINGLASS

Ex MALVA 1919 won 3 races, d. of *King Salmon E & †His Grace
- *†CHARLES O'MALLEY 1907 p. 64
  - Desmond: St. Simon, L'Abbesse De Jouarre O
  - Goody Two Shoes: ISINGLASS, Sandal
- Wild Arum 1911
  - Robert Le Diable: Ayrshire, Rose Bay
  - Marliacea: Martagon, Flitters

---

by BLENHEIM

ex DELLEANA 1925 (It.2,1) & 6 other races
- †CLARISSIMUS 1913 (exp. Fr.'21) 2,Ch. sired also the stayer Nino and the dams of *†Brantome AT, Crudite GP En Fraude FO, *†Pharis II FD,GP & Sirtam.
  - Radium: Bend Or, Taia
  - Quintessence 1: St. Frusquin, Margarine
- DUCCIA DI BUONINSEGNA 1920 Bred in Ireland, exp. Italy 1921. won It. 1000.
  - Bridge of Earn: Cyllene, Santa Brigida
  - Dutch Mary: William The Third, Pretty Polly p. 73

By *†DONATELLO II 1934 (It.D) — SWYNFORD

Ex AURORA 1936 see p. 17
- *†HYPERION 1930 D,L
  - Gainsborough: Bayardo to Eclipse, Rosedrop O
  - Selene: Chaucer, Serenissima
- ROSE RED 1924 p. 133
  - SWYNFORD: John O'Gaunt, Canterbury Pilgrim O
  - Marchetta 4th d. of *†My Babu 2: Marco, Hettie Sorrel

ISONOMY

Isinglass
John O'Gaunt
Swynford
Blandford —— Bahram see later
Blenheim
Donatello II

1954 to 1974 *†CREPELLO 2,D see p. 128

By *†DONATELLO II 1934 (It.D)
- *†BLENHEIM 1928 D
  - Blandford [Swynford / Blanche]
  - Malva [Charles O'Malley / Wild Arum]
- DELLEANA 1925
  - Clarissimus [Radium / Quintessence 1]
  - Duccia Di Buoninsegna [Bridge of Earn / Dutch Mary]

Ex CREPUSCULE 1948 Won 1 small 10f race dam of Honeylight 1 & Twilight Alley GC
- *†MIEUXCE 1933 FD,GP Imp.UK 1936 won also Lupin & 2 other races at 3y. Sired also Commotion O (dam of *Combat) & the dams of Margueritte Vernaut Ch. & *Shikampur.
  - Massine (appx. A) [Consols to Eclipse / Mauri]
  - L'Olivete [Opott / Jonicole]
- Red Sunset
  - Solario [Gainsborough / Sun Worship]
  - Dulce II [Asterus / Dorina FO p. 127]

1931 *†BRANTOME AT (F2, Lupin, FL, Cadran)

Plus 7 other races including the Grand Criterium as a 2yo. He was beaten only twice. He became a rogue. He sired Pensbury GP. Vieux Manoir (next), †Breughel and the dams of Tropique E, Timandra FO, (F1) and Guersant (F2). One wonders how long this line will survive without the injection of some sprinting blood.

By *†BLANDFORD 1919
- Radium [Bend OR / Taia]
- Quintessence 1 [St. Frusquin / Margarine]

Ex VITAMINE 1924 dam of Crudite 1932 GP
- *†CLARISSIMUS 1913 2,Ch.
  - Radium
  - Quintessence 1
- Viridiflora
  - Sans Souci II [Le Roi Soleil / Sanctimony]
  - Rose Nini d. of Rose Verte FO [Le Sancy / Rosewood 1883 exp. France 1900]

1947 *VIEUX MANOIR GP

and two other 12½f races. Sired San Roman GP, *Le Haar, *Mourne and *Val De Loir (now unfortunately dead after having sired Tennyson GP and Val De L'Orne FD). All good stayers; would that they and their stallion progeny could be mated with mares with speed close up in their pedigrees.

By *†BRANTOME 1931 AT

Ex VIELLE MAISON 1936 p. 88.
- †FINGLAS 1923 p. 24
  - Bruleur [Chouberski to Herod / Basse Terre]
  - Fair Simone [Farman / Simone]
- Vielle Canaille
  - Zionist [Spearmint / Judea]
  - Ficelle [La Farina / Quenouille FO p. 88]

ISONOMY

Isinglass
John O'Gaunt
Swynford
Blandford

1932 to 1956 — *†BAHRAM 2,D,L, see p. 47

By *†BLANDFORD 1919	*†SWYNFORD 1907 L,E	John O'Gaunt	Isinglass
			La Fleche
		Canterbury Pilgrim	Tristan
			Pilgrimage
	BLANCHE	White Eagle	Gallinule
			Merry Gal
		Black Cherry	Bendigo
			Black Duchess
Ex FRIAR'S DAUGHTER 1921	*†FRIAR MARCUS 1912 p. 47	Cicero	Cyllene to Eclipse
			Gas
		Prim Nun	Persimmon
			Nunsuch
	GARRON LASS 1917	Roseland	William The Third
			Electric Rose
		Concertina	St. Simon
			Comic Song

1940 to 1964 — *†PERSIAN GULF see p. 140

By *†BAHRAM 1932 2,D,L	*†BACHELOR'S DOUBLE 1906 p. 79	Tredennis	Kendal
			St. Marguerite 1
		Lady Bawn	Le Noir
			Milady
Ex DOUBLE LIFE 1926 p. 139	Saint Joan 1918	Willbrook	Grebe
			Nora Gough
		Flo Desmond	Desmond
			Flighty Flo

THE DARLEY ARABIAN
to
Eclipse
to
Pot.8.os
to
Whalebone
to
Camel
to
Touchstone
to
HAMPTON

1893 to 1907 *BAY RONALD

Won the Hardwicke Stakes as a 4yo and the City and Suburban h'cp. when he was 5, but as he was only carrying 8 stone it looks as if he had deteriorated. However, he sired the great Bayardo (next) before being sold to France in 1905 where he died two years later. His other sons were *Macdonald II(FL), Bay Cherry, Rousay and *†Dark Ronald (p.69, App.C, D). He played a more important part as a sire than his racing career would suggest, for he sired the mare Rondeau (p.149, App. C) the dam of *†Teddy FD, (p.129, 149, App. C & D).

By *†HAMPTON 1872 p. 7	*†LORD CLIFDEN 1860 L	Newminster	Touchstone
			Beeswing
		The Slave	Melbourne
			Volley
	LADY LANGDEN 1868	Kettledrum	Rataplan
			Hybla
		Haricot	Lanercost
			Queen Mary
Ex BLACK DUCHESS 1886 p. 35	*†GALLIARD 1880 2	Galopin	Vedette to Eclipse
			The Flying Duchess
		Mavis	Macaroni
			Merlette
	Black Corrie 1879	Sterling	Oxford
			Whisper
		b mare 1861	Wild Dayrell
			Lady Lurewell

1906 to 1917 *†BAYARDO L,GC,E,Ch

and 18 other races, see p.30, App. D.

By *BAY RONALD 1893			GALOPIN
			STERLING
Ex GALICIA 1898 won once at 2y	*†GALOPIN 1872 D	Vedette	Voltigeur to Eclipse
			Mrs Ridgeway
		The Flying Duchess	The Flying Dutchman
			Merope
	ISOLETTA 1891 p. 29	Isonomy	STERLING
			Isola Bella
		Lady Muncaster	Muncaster
			Blue Light

1915 to 1945 *†GAINSBOROUGH 2,D,L,GC
see p. 83

By *†BAYARDO 1906 L,GC,E,Ch.		GALOPIN	
Ex ROSEDROP 1907 O, Exp. USA '22	*†ST. FRUSQUIN 1893 2,D, p. 39	St. Simon	GALOPIN
			St. Angela
		Isabel	Plebeian
			Parma
	Rosaline 1901 Exp. Arg. '10	Trenton	Musket
			Frailty
		Rosalys 1894 exp. Ger. 1906	Bend Or
			Rosa May

HAMPTON

Bay Ronald
Bayardo
Gainsborough

1930 to 1960 *†HYPERION D,L
see p. 83

1938 to 1965 *†OWEN TUDOR D,GC

Both wartime substitutes run at Newmarket, despite having gone to stud during the war he does not appear to have lacked good mares. Of his first crop, Tudor Minstrel won the 2000, next year Abernant was 2nd in it; both of these very good horses were short runners, Tudor Minstrel won no further than a mile and Abernant never more than 7f. Owen Tudor also got Elpenor GC (exp. Brazil) and Right Royal V (next) who just has a chance of keeping this branch of Hyperion at classic level.

1958 to 1973 *RIGHT ROYAL V FD,GVI,(F2)

and 5 other good races in his native France including the Grand Criterium for 2yos. He sired Prince Regent IRD and the dam of Ragstone GC.

1950 to 1974 *AUREOLE GVI

Vienna next page

Sire / Dam	Grandparents	Great-grandparents	Great-great-grandparents
By *†GAINSBOROUGH 1915 2,D,L,GC			GALOPIN
			ST. SIMON
Ex SELENE 1919	*†CHAUCER 1900 p.38, 152, 156, 158, 167, App. D.	ST. SIMON	GALOPIN
			St. Angela
		Canterbury Pilgrim	Tristan
			Pilgrimage
	SERENISSIMA 1913	Minoru	Cyllene
			Mother Siegel
		Gondolette	Loved One
			Dongola
By *†HYPERION 1930 D,L		CHAUCER	
Ex MARY TUDOR II 1930 (F1) imp. UK ex France	*†PHAROS 1920 Ch. p. 56	Phalaris	Polymelus to Eclipse
			Bromus
		Scapa Flow	CHAUCER
			Anchora
	ANNA BOLENA 1920 (F1)	Teddy	Ajax
			Rondeau
		Queen Elizabeth II exp. France 1913	Wargrave
			New Guinea
By *†OWEN TUDOR 1938 D,GC			
Ex BASTIA 1951 did not win imp UK '59	*†VICTRIX 1934 (FL) Sired also Avenger GP & the dam of Balto GP, GC	Kantar	Alcantara II to Eclipse
			Karabe
		Victory	Swynford
			Lineage
	BarberyBush 1934	Ksar	Bruleur
			Kizil Kourgan
		Pervencheres	Maboul II
			Poet's Star exp. France 1919
By *†HYPERION 1930 D,L			
Ex ANGELOLA 1945 won 4 races p. 142, App. D.	*†DONATELLO II 1934 (It.D) p.74	Blenheim	Blandford to Eclipse
			Malva
		Delleana	Clarissimus
			Duccia Di Buoninsegna
	FEOLA 1933 p. 80	Friar Marcus	Cicero
			Prim Nun
		Aloe	Son in Law
			Alope p. 80, App. D.

HAMPTON

Bay Ronald
Bayardo
Gainsborough
Hyperion
Aureole

1957 *ST. PADDY D,L,E

and 6 other races including the Hardwicke Stakes, see p.110 Sire of Connaught E. and of the dam of Dunfermline O,L.

By *†AUREOLE 1950 GVI			
Ex EDIE KELLY 1950 won 1 5f race at 3y	*†BOIS ROUSSEL 1935 D p. 46	Vatout	Prince Chimay to Eclipse
			Vasthi
		Plucky Liege	Spearmint
			Concertina
	Caerlissa 1935	Caerleon	Phalaris
			Canyon 1
		Sister Sarah	Abbots Trace
			Sarita p. 112

1957 *VIENNA see p. 75

By *AUREOLE GVI			
Ex TURKISH BLOOD 1944 won twice 5f at 2y	TURKHAN 1937 L, (Ir.D) exp. France 1952	Bahram	Blandford to Eclipse
			Friar's Daughter
		Theresina	Diophon
			Teresina
	Rusk 1935	Manna	Phalaris
			Waffles
		Baby Polly	Spearmint
			Pretty Polly p.32, 73-75, 169, App. D.

1965 *VAGUELY NOBLE AT

At stud in USA, see p.118, App. D.

By *VIENNA 1957		<u>HYPERION</u>	<u>BAHRAM</u>
Ex NOBLE LASSIE p. 118	*†NEARCO 1935 GP	Pharos	Phalaris to Eclipse
			Scapa Flow
		Nogara	Havresac II
			Catnip
	Belle Sauvage 1949	Big Game	<u>BAHRAM</u>
			Friar's Daughter
		Tropical Sun	<u>HYPERION</u>
			Brulette O p. 118

THE GODOLPHIN ARABIAN
to
Matchem
to
Trumpeter
to
Comus
to
Melbourne
to
WEST AUSTRALIAN

Solon — Australian see later

Solon
Barcaldine

1892 *MARCO

1903 to 1917 *†MARCOVIL

A handicapper like his sire, won 2 races, went lame but was patched up to win the Cambridgeshire as a 5yo car. 7.11. Besides Hurry On (next) the only animal of consequence he sired was Golden Orb (p. 12) the maternal grandsire of Bellacose.

1913 to 1936 *†HURRY ON L

and 5 other races; unbeaten, see p.139, App. D.

By *†BARCALDINE 1878 a large Irish stayer, unbeaten	SOLON 1861 Irish	West Australian	Melbourne
			Mowerina 1867
		'DARLING'S DAM'	Birdcatcher
			b Mare 1846 Hetman Platoff
	Ballyroe 1872	Belladrum	Stockwell
			Catherine Hayes O
		Bon Accord	Adventurer
			'DARLING'S DAM'
Ex NOVITIATE 1882 won 1 small race	*†HERMIT 1864 D p. 11	Newminster	Touchstone to Eclipse
			Beeswing
		Seclusion	Tadmor
			Miss Sellon
	Retty 1865	Lambton	The Cure
			Elphine
		Fern	Fern Hill
			mare 1848 by Beiram, d. of 1859 Grd. Nat. wnr. Half-Caste

By *MARCO 1892		HERMIT	
Ex LADY VILLIKINS 1895 did not run 2nd dam of *Radames.	HAGIOSCOPE 1878 A stayer, ran 42 times, won 13 races.	Speculum	Vedette to Eclipse
			Doralice
		Sophia	Macaroni
			Zelle
	Dinah 1871	HERMIT	
		The Ratcatcher's daughter, 2nd d. of †Amphion Ch.	Seclusion
			Rataplan
			Lady Alice

By *†MARCOVIL 1903			HERMIT
			HERMIT
Ex TOUTE SUITE 1904 did not run	*†SANFOIN 1887 D p. 151	Springfield	St. Albans to Eclipse
			Viridis
		Sanda	Wenlock
			Sandal
	Star 1887	Thurio	Cremorne
			Verona
		Meteor 1867 2nd d. of *Sheen	Thunderbolt
			Duty

WEST AUSTRALIAN

Australian see later

Solon
Barcaldine
Marco
Marcovil
Hurry On

1933 to 1937 — *†PRECIPITATION GC see p. 140

Parents	Grandparents	Great-grandparents	Great-great-grandparents
By *†HURRY ON 1913 L			
	*†BACHELOR'S DOUBLE 1906 p.79, 165, 171, App. D.	Tredennis	Kendal to Eclipse
			St. Marguerite 1
		Lady Bawn	Le Noir
			Milady
Ex DOUBLE LIFE p. 139	Saint Joan 1918	Willbrook	Grebe
			Norah Gough
		Flo Desmond	Desmond
			Flighty Flo

1956 — *SHESHOON GC see p. 24

Parents	Grandparents	Great-grandparents	Great-great-grandparents
By *†PRECIPITATION 1933 GC			
	*†NEARCO 1935 GP	Pharos	Phalaris
			Scapa Flow
		Nogara	Havresac II
			Catnip
Ex NOORANI 1950 p.24 d. of Charlottesville 1957 FD,GP	Empire Glory 1930	Singapore	Gainsborough
			Tetrabazia
		Sky Glory	Sky Rocket
			Simone

WEST AUSTRALIAN

Australian Spendthrift

ALL U.S.A.

1893 HASTINGS

Won Belmont Park Stakes

1905 *FAIR PLAY see p. 12

1917 to 1947 *†MAN O'WAR

Won Belmont & Preakness Stakes and 18 other big races, he was beaten only once in 21 starts. He became Champion Sire of USA and, on his death aged 30, he was entombed at Lexington and an immense statue erected in his honour. So far as European racing is concerned he sired: – *†War Admiral (next), War Relic, American Flag, Battleship (Grand National), and the dam of Blue Swords.

By SPENDTHRIFT 1876	AUSTRALIAN 1858 Exp. USA 1869	West Australian	Melbourne
			Mowerina 1843
		Emilia	Y. Emilius
			Persian
	Aerolite 1861	Lexington	Boston
			Alice Carneal
		Florine	Glencoe
			Melody
Ex CINDERELLA 1885 exp. From UK to USA 1886	Tomahawk died 1884 Or (Blue Ruin US)	King Tom	Harkaway to Eclipse
			Pocahontas
		Mincemeat O	Sweetmeat
			Hybla
	Manna 1874	Brown Bread	Weatherbit
			Brown Agnes
		Tartlet	Birdcatcher
			b mare 1848 by Don John

By HASTINGS 1893	*†BEND OR 1877 D,Ch.	Doncaster	Stockwell to Eclipse
			Marigold
		Rouge Rose	Thormanby
			Ellen Horne
Ex FAIRY GOLD exp. from UK to USA 1903	DAME MASHAM 1889	Galliard	Galopin
			Mavis
		Pauline	Hermit
			Lady Masham p.111

By *FAIR PLAY 1905	*†ROCK SAND 1900 2,D,L exp. USA 1906 & to France 1912. Inbred to Stockwell 3 times at 4th rem. sired Quelle Est Belle FO, *†Tracery L,E, Ch., Rock Flint, Trap Rock & Friar Rock	Sainfoin	Springfield to Eclipse
			Viridis
		Roquebrune 1893 exp.Bel. 1905	St. Simon
			St. Marguerite
Ex MAHUBAH	MERRY TOKEN 1891 exp. USA 1902	Merry Hampton	Hampton
			Doll Tearsheet
		Mizpah	Macgregor
			b mare 1863 by Underhand

WEST AUSTRALIAN

Australian
Spendthrift
Hastings
Fair Play
Man O'War

1934 — *†WAR ADMIRAL

Won Preakness, Belmont and Kentucky Derby 21 races out of 26 starts, unplaced only once. Sired †Great Captain (sire of dam of Peleid L), and the dams of *Never Say Die D,L & Gun Bow (sire of Pistol Packer FO), also sired the mare Searching (won 25 races and was 2nd dam of Allez France FO AT).

- By *†MAN O'WAR 1917
- Ex BRUSH UP 1929 bred 6 winners
  - SWEEP 1907
    - Ben Brush
      - Bramble to Eclipse
      - Roseville
    - Pink Domino
      - Domino
      - Belle Rose
  - ANNETTE K 1921 exp. in utero
    - Harry of Hereford
      - John O'Gaunt
      - Canterbury Pilgrim
    - Bathing Girl exp. USA 1920
      - Spearmint
      - Summer Girl

1938 — WAR RELIC

Won 9 races at 3y; has entered into European classics so far only through his son Relic (next)

- By *†MAN O'WAR 1917 — — — — — — — FAIRY GOLD / ROCKSAND
- Ex FRIAR'S CARSE 1923
  - *FRIAR ROCK 1913 (Belmont), Sired Pilate.
    - ROCKSAND
      - Sainfoin to Eclipse
      - Roquebrune
    - FAIRY GOLD
      - Bend Or
      - Dame Masham
  - Problem 1913
    - Superman
      - Commando
      - Anomaly
    - Query
      - Voter
      - Quesal

1945 to 1970 — *†RELIC

Won 5 races up to 7f; imported to France from USA in 1950 and to UK in 1957. Sired Pieces Of Eight 1963 E,Ch. on whom it looks as if the responsibility rests for keeping this Man O'War stirp alive in European classic breeding. But, whatever happens, Relic will be revered I hope for a long time for his daughter Relance III (p.67) dam of *Match III GVI, Reliance FD,GP and *Relko D.

- By WAR RELIC 1938
  - FAIRY GOLD
  - ROCKSAND
  - ROCKSAND
  - — — — — — — — — — — FAIRY GOLD
- Ex BRIDAL COLORS 1931
  - *†BLACK TONEY 1911 Sired Black Servant, Balladier, Bimelech (Bel, Preak) & also the dam of Great Captain.
    - Peter Pan
      - Commando to Eclipse
      - Cinderella
    - Belgravia
      - Ben Brush
      - Bonnie Gal
  - VAILA 1911 exp. USA 1915
    - Fariman
      - Gallinule
      - Bellingorm
    - Padilla
      - Macheath
      - Padua

THE BYERLEY TURK
to
Herod
to
Selim
to
GARDEFEU

1902 †CHOUBERSKI

Won the only race he ran in, then split a pastern. His only classic son was Bruleur (next), his daughters bred *Biribi AT,(FL), Kantar AT and Kandy 1.

By *GARDEFEU 1895 FD (lupin)	CAMBYSE 1884	Androcles	Dollar
			Alabama
		Cambyse	Plutus
			Campeche FO
	Bougie 1887	Bruce	See Saw
			Carinne
		La Lumiere	The Heir of Lynne
			Grand M'slle
Ex CAMPANOULE 1891	*THE BARD 1883 exp. France 1887 Won Goodwood & Doncaster Cups. Alleged to have been the only horse to have made Ormonde gallop.	Petrarch	Lord Clifden to Eclipse
			Laura
		Magdalene	Syrian
			My Mary
	St. Lucia 1880 exp. France 1883	Rosicrucian	Beadsman
			Madame Eglantine
		Rose of Tralee	Knowsley
			Vimiera

1910 *†BRULEUR GP,(FL) see p. 117

By †CHOUBERSKI 1902	†OMNIUM II 1892 FD	Upas	Dollar to Herod
			Rosemary
		Bluette	Wellingtonia
			Blue Serge
Ex BASSE TERRE 1899 4th d. of Pointe a Pitre FO.	Bijou 1890 exp. France 93	St. Gatien	The Rover
			St. Editha
		Thora	Doncaster
			Freia

1918 *†KSAR FD,AT(twice)

And other big races in France. It is said that in appearance he resembled Omnium II and that may well have been the case considering that he is inbred to that animal to the point of incest. He sired Ukrainia FO, *†Tourbillon FD (next), Thor FD, Le Ksar 2, †Amfortas, Muzio and Castel Fusano. His daughters bred †Le Pacha FD,GP,AT,(FL), Kingsway 2, Stratonice FO and *Majano. Ksar was exp. USA 1935.

By *†BRULEUR 1910 GP		†OMNIUM II	
	†OMNIUM II	Upas	Dollar to Herod
			Rosemary
		Bluette	Wellingtonia
			Blue Serge
Ex KIZIL KOURGAN 1899 FO,GP,(F1) 2nd d. of Kandy 1.	KASBAH 1892 FO 3rd d. of Kantar AT	Vigilant	Vermouth
			Virgule
		Katia	Guy Dayrell
			Keepsake

GARDEFEU

Chouberski
Bruleur
Ksar

1928 *†TOURBILLON FD
see p. 90

1937 to 1958 *†DJEBEL 2,AT,(F2)

and 12 other races, winning 7 off the reel as a 5 yo. Small and tough, he was kept clear of the vicissitudes of the German occupation and went to stud in 1943. 3 times champion sire in France, his first crop included *Clarion (next), Le Lavandou (sire of Le Levanstell), Montenica FO and the stayer Arbar GC,(Cadran). Thereafter the invasion of UK continued with *†My Babu 2, Djeddah E,Ch., Galcador D, Argur E and Hugh Lupus Ch. (Ir.2). He also got *Djefou, Apollonia FO,(F1) and, one would think as an experiment, the filly Coronation V AT, (=F1) who only had 3 grandparents, her dam being by Tourbillon.

1944 *CLARION
see p. 41

Whose best son and just possibly his linear successor *†Klairon (F2) has sired Lorenzaccio 1965 Ch. with whom, it seems, the matter rests.

1952 to 1967 *HUGH LUPUS CH.(Ir.2)

Foaled France imp. UK '55, also won 6 other races up to 12f. (Hardwicke Stakes). He died young but not before siring *Hethersett L (next) and Pourparler 1.

By *†KSAR 1918 FD,AT (twice)		OMNIUM II	OMNIUM II
			St. Simon to Eclipse
			Satirical
	†DURBAR II 1911 D exp. USA 1926	Rabelais	The Meddler
		Armenia	Urania
Ex DURBAN 1918			Candlemas
		Irish Lad	Arrowgrass
	BANSHEE (F1) 1910		Hamburg
		Frizette	Ondulee exp. USA 1903

By *†TOURBILLON 1928 FD			Bay Ronald
	*†GAY CRUSADER 1914 2,D,L,GC,Ch, p. 162	Bayardo	Galicia
			Beppo
Ex LOIKA 1926 won once, dam of Hierocles and 5 other winners.		Gay Laura	Galleotia 1
			Ajax
		Teddy	Rondeau
	Coeur a Coeur 1921 did not win bred 5 wnrs.		Ayrshire
		Ballantrae exp. USA 1904	Abeyance

By *†DJEBEL 1937 2,AT		GAY CRUSADER	Polymelus
		Phalaris	Bromus
	*†COLORADO 1923 2,E p. 54		Chaucer
		Canyon 1	Glasalt
Ex COLUMBA 1930 did not win			Bayardo
		GAY CRUSADER	Gay Laura
	Gay Bird 1923 Exp. USA 1934		St. Frusquin
		Popinjay	Chelandry 1 p. 39

By *†DJEBEL 2, AT	*†TOURBILLON 1928 FD		BRULEUR
			Ksar
	*†GOYA II 1934 p. 123	TOURBILLON	Durban
			Sardanapale
Ex SAKOUNTALA 1942 won 1 good race imp. UK 1952		Zariba	St. Lucre
			Chouberski
	SAMOS 1932 AT bred 5 wnrs.	BRULEUR	Basse Terre
			Nimbus 1910
		Samya bred 8 wnrs.	Sapience, sis. to Sourbier FD

GARDEFEU

Chouberski
Bruleur
Ksar
Tourbillon
Djebel
Hugh Lupus

1959 to 1966 *HETHERSETT L,

Fell in the Derby when favourite and unfortunately died 4 years later having sired: –

1966 *BLAKENEY D

and 2nd by a length in the GC next year. He stands now at the National Stud and is getting good mares; one of the first he served has produced the Oaks winner Juliette Marny. It could just happen that he will be the one to keep the flag of the distant Herod flying in Europe. Since when he has sired Juliette's brother, Julio Mariner L.

By *HUGH LUPUS 1952 Ch.		TOURBILLON	TOURBILLON
Ex BRIDE ELECT 1952 p. 116	*†BIG GAME 1939 2, Ch. p. 82	Bahram	Blandford
			Friar's Daughter
		Myrobella	Tetratema
			Dolabella
	Netherton Maid 1944	Nearco	Pharos
			Nogara
		Phase, d. of Neasham Belle O & Narrator Ch.	Windsor Lad
			Lost Soul

By *HETHERSETT 1959 L			NEARCO
Ex WINDMILL GIRL 1961 won 1 good race, 2nd in Oaks, later to become dam of Morston D.	*†HORNBEAM 1953 p.	Hyperion	Gainsborough
			Selene
		Thicket	Nasrullah
			Thorn Wood
	Chorus Beauty 1952	Chanteur II	Chateau Bouscaut
			La Diva
		Neberna	NEARCO
			Springtime p. 132

APPENDIX A
Other contributing stallions

Admiral's Walk 1936, St. James' Palace Stakes; by Hyperion ex a Roi Herode mare.

***†Alcantara II** 1908 FD (Lup.), by Perth ex Toison d'Or by Le Sancy; sired Flowershop FO, Kefain GP, Aquatinte FO, Kantar AT, Kandy 1, *Pinceau, Indus (F2), & the dams of Thor FD, *Fastnet & Brumeux.

***†Alizier** 1947, stayer, 9 races in France; by Telerifique ex a Coronach mare; sire also of Tapalque FD & the dam of Goodly FD.

Amadis 1906, Donc. Cup, Gold Vase etc.; by Love Wisely ex a Ladas mare.

Arion 1915, mid. dist. h'cpr; by Valens ex mare by Galloping Lad.

Arjaman 1930, German; by Herold ex mare by Fervor, inbred to Ard Patrick at 3rd & 4th removes.

Badajoz 1907, French stayer; by Gost ex a Raeburn mare; sired also *†Epinard.

Balladier 1933, USA; by Black Toney ex mare by North Star III.

Balthazar 1937, French, won up to 9f; by Epinard ex mare by Amant de Coeur.

***Beau-Pere** 1927, by Son In Law ex Cinna 1 by Polymelus; champion sire twice N.Z. & 3 times in Australia, then to USA where also he was a success, inbred to Hampton at the 4th remove.

Beau Prince II 1952, won 8 French races up to 10f inc. Prix Ganay; by Prince Chevalier ex mare by Black Devil. Exp. Fr. '52.

Bellacose 1932, sprinter, won Nunthorpe & the July Cup twice, 12 wins in all; by Sir Cosmo ex Golden Orb mare.

***†Biribi** 1923 AT (Lup. FL); by Rabelais ex Chouberski mare, sired also Perruche Bleu FO & †Le Pacha FD GP AT (FL).

***†Black Devil** 1931, Donc. Cup & 3 other races; by Sir Galahad III ex Ambassador mare, inbred to Ajax at 3rd & to Bay Ronald the 4th removes; sired also Chamant FD & Beau Prince II.

***†Blue Skies** 1929, won 2 mid. dist. races in France; by Blandford ex Blue Pill by Sans Souci II; sired also the dams of *†Chanteur II & *Turmoil.

Bold Archer 1924, won 3 races at 2y inc. Gimcrack; by Phalaris ex the prolific Miss Matty by Marcovil, inbred to St. Simon at the 3rd & 4th removes.

Brown Bud 1924 USA; by Brown Prince (exp. 1920) ex Mare by Myram, inbred to St. Simon at the 4th remove.

†By George 1911, a great 2yo; at stud Ireland 9 years, exp. Canada 1926; by Lally ex Queen's Holiday by Royal Hampton, inbred to Galopin at the 4th rem.

Caligula 1927, French; by Town Guard ex Catherine Swynford (exp. to France 1920) by Swynford; sired also the dam of Cosmos.

***Cannobie** 1913, stayer, exp. France 1920; by Polymelus ex La Roche O by St. Simon, sired also Cranach (1923) & Pitchoury. Exp. Fr. '20.

Caruso 1927, USA; by Polymelian ex a Harmonicum mare, inbred to Amphion at the 4th remove.

Casquetts 1913, by Captivation ex Cassis by Morion, gddam of *†Pharis II FD, GP (p s8) & gt gddam of Elpenor GC. Exp. Fr. '21.

Chut 1909, French; by Ajax ex Roquette (dam of Dagor FD) by Chaleureux; inbred to Galopin at the 4th remove; sired also Quoi FO.

***†Count Fleet** 1940, USA, won Belmont Stakes & Kentucky D; by Reigh Count ex mare by Haste; inbred to St. Frusquin at the 4th remove.

***†Coup de Lyon** 1930, won 5 races up to 10f; by Winalot ex a Sunstar mare, sired also the granddam of both Pourparler 1 & Fleet 1.

Crafty Admiral 1948, USA, won 18 races; by Fighting Fox ex a War Admiral mare; sired also the granddam of Nonoalco 2.

†Cranach 1938, in Germany 1940-43, no racing record; by Coronach ex a Blandford mare.

Crapom 1930 AT & 6 other races; by Cranach (1923) ex Adam mare: inbred to Cyllene at the 4th remove. Exp. Ger. '40 from It.

Cylgad 1909, mid. dist.; by Cyllene ex the staying mare Gadfly by Hampton; sired also Tangiers GC & the dams of †Pappageno II & *Merry Boy. Exp. Fr. '22.

Democratic 1952, sprinter; by Denturius ex Signal Light mare. Exp. Arg. '66.

†Denturius 1937, won eight 5f sprints, 7 of them in Ireland; by Gold Bridge ex Lemberg mare; inbred to Orby at the 4th remove; sired Democratic & the dam of Abermaid 1.

***†Diligence** 1919, mid. dist.: by Hurry On ex Radium mare. Exp. Rus. '35.

Diolite 1927 2, by Diophon ex Rocksand mare. Exp. Jap. '35.

Ellangowan 1920 2 Ch. & St. James' Palace Stakes; by Lemberg ex Sunstar mare; inbred to Isonomy at 4th rem.

***†Epinard** 1920, very fast, won Stewards' Cup + 11 races in France; by Badajoz ex a (USA) Rocksand mare; inbred to St. Simon at the 4th remove; sired also *†Rodosto 2 (F2) & Balthazar.

Ethnarch 1922, won 4 good races; by The Tetrarch ex William The Third mare; sired also the dams of *Khaled & Piping Rock.

Fairy King 1910, won 11 races; by Desmond ex Oberon mare; inbred to Galopin at the 3rd remove. Exp. Ger. '22.

***†Fantastic** 1933 won FL, Cadran & 6 other races; by Aethelstan ex Durbar II mare; sired also the dam of Free Man.

†Farasi 1903, won at 2y at Newmarket, at stud in Ireland (19gns later reduced to 5); by Desmond ex mare by Morgan. see p.15.

***Felicitation** 1930 GC & 5 other good races; by Colorado ex Cantilever mare; inbred to St. Simon at 4th rem. Exp. Braz. '44.

***†Felstead** 1925 D, by Spion Kop ex Lemberg mare; inbred to Carbine at the 3rd & 4th removes; sired also Rockfel 1 0 Ch., Steady Aim O & the dam of *Rockefella.

***Firdaussi** 1929 L & 7 other races; by Pharos ex Brownhylda O by Stedfast; inbred to Chaucer at the 3rd rem.; sired also †Escamillo. Exp. Fr. '34, later to Rumania.

***†Fiterari** 1924 GP (F2) (FL); by Sardanapale ex Bachelor's Double mare; inbred to St. Simon at the 4th rem.

†Flush Royal 1945, stayer, ran 51 times, placed 24 & won 18 including the Cesarewitch (car. 8.13), raced 8 seasons; by Majano ex Olibreus mare.

***Gallant Fox** 1927, won Kentucky D, Belmont & Preakness Stks. by Sir Galahad III ex Marguerite (dam of Petee Wrack) by Celt; lived to be 27 and sired also Flares GC Ch. & Omaha (USA triple Crown like his sire).

***Galloper Light** 1916 GP; by Sunstar ex St. Frusquin mare; sired also Beam O & Light Brocade O.

Galloping Simon 1904, did not run; by Melton ex Simena by St. Simon; inbred to Galopin at 3rd & 4th & to Stockwell at 4th removes.

***Gold Bridge** 1929, won nine 5f races inc. Nunthorpe & King's Stand Stakes; by Golden Boss ex a Diadumenos mare; inbred to Orby at the 3rd remove; sired also. †Denturius & Vilmorin. Fld. Fr., imp. '32.

Go To Bed 1903, French; by Perth ex Krakatoa mare; sired also the dams of Pitchoury & Davout (F2).

Graf Isolani 1926 (Ger. D); by Graf Ferry ex Majestic mare, inbred to St. Simon at the 4th remove.

***†Hasty Road** 1951, USA, winner of 14 races inc. Preakness Stks; by Roman ex Traffic Court by Discovery.

He 1915, good mid. dist. horse, by Santoi ex She by Cyllene.

Hierocles 1939, won 6 good mid. dist. races in France, exp. USA 1947; by Abjer ex the dam of Djebel, Loika by Gay Crusader; inbred to Teddy at the 3rd remove.

†His Grace 1933, won 6 races inc. Coronation Cup; brother to Bleinheim, by Blandford ex Malva by Charles O'Malley; inbred to Isinglass at the 4th rem. Sired also the dam of Noblesse O.

Hoist The Flag 1968 (USA) by Tom Rolfe (Preak) ex a War Admiral mare. Sired ALLEGED AT (twice).

Horus 1926, won 4 good mid. dist. races; by Papyrus ex Lady Peregrine by White Eagle (dam of Flamingo 2); inbred to St. Simon at the 4th remove.

Jock II 1936, French, did not run until 1940 & '41, stayer, won 5 races; by Asterus ex Naic by Gainsborough; inbred to Bay Ronald at the 4th remove; sired also *†Sunny Boy.

†**Johnstown** 1936, won Kentucky D, Belmont & 12 other races, by Jamestown ex Sir Galahad III mare; inbred to Ajax at the 4th remove; sire also of Nashua.

Junior 1909, stayer, won 10 races; by Symington ex Scylla by Eager; inbred to Galopin at the 4th remove; sired also the dam of Plack 1.

Legatee 1923, mid. dist.; by Gay Crusader ex Amadis mare, inbred to Galopin at the 4th remove.

Le Volcan 1941, sprinter, by Tourbillon ex Banstar mare, inbred to Durbar II at 3rd & 4th removes; sired also the dam of Midnight Sun.

Loliondo 1937, French, mid. dist.; by Badruddin ex Tetrameter mare; inbred to The Tetrarch at the 3rd remove.

Lycaon 1908, stayer, exp. Germany; by Cyllene ex Hampton mare.

***The Macnab** 1926, miler, by Abbots Trace ex Sunstar mare; sired also Vigilance FO.

*†**Massine** 1920 GC AT, by Consols ex Mauri by Ajax; inbred to St. Simon at the 4th remove; sired also Strip The Willow FD, GP, Feerie FO, *†Maravedis & the dam of *†Fontenay.

***Mr. Jinks** 1926 2 & eight other races (av. dist. 7.7f); by Tetratema ex Lemberg mare; inbred to Bona Vista at the 4th rem.

†**Nepenthe** 1938, French stayer, by Plassy ex Isard II mare; sired also the dam of Amber FD.

***Nino** 1923, French stayer, won 9 races; by Clarissimus ex Azalee by Ajax; sired also *Canot.

Northeast 1905 GP, by Perth ex Le Hardy mare.

*†**Nosca** 1937, French miler, by Abjer ex Tourbillon mare; sired also Cosmos and the dam of Pandofell GC.

Orpen 1928, stayer, won 5 races inc. Hardwicke Stks.; by Solario ex Harpy by Swynford.

Orpheus 1917 Ch. twice & 9 other races; by Orby ex Electra by Eager; sired also Ormenos.

Over There II 1916, USA, by Spearmint ex Sundridge mare.

***Pharamond II** 1925, good 2yo, leading sire USA; by Phalaris ex Selene (Hyperion's dam) by Chaucer; inbred to both Cyllene & St. Simon at the 3rd & 4th removes. Exp. USA '28.

Picton 1903, by Orvieto ex Isonomy mare, inbred to Stockwell & Thormanby at the 4th remove.

Pilade 1930, won Italian D; by Captain Cuttle ex Woolwinder mare.

Pot o'Luck 1942, USA stayer, won 10 races; by Chance Play ex Wildair mare; inbred to Peter Pan at the 4th rem.

Quai d'Orsay 1931, won 12 mid. dist. races; by Town Guard ex Gorgos mare; inbred to St. Simon at the 4th remove.

Rochester 1907, won up to a mile; by Rocksand ex Sir Visto mare.

*†**Rodosto** 1930 2 (F2) & seven other races; by Epinard ex Ramondie by Neil Gow; sired also the dam of Douve FO. Exp. Arg.

Sandwich 1928 L, by Sansovino ex Waffles (dam of Manna) by Buckwheat; inbred to St. Simon at the 3rd & 4th and to Islinglass at the 4th removes. Exp. Braz. '41.

Scarlet Tiger 1930, mid. dist.; by Colorado ex Trilogy by Son In Law; inbred to Cyllene & St. Simon at the 4th remove. Exp. Arg. '41.

Scotland 1892, bro. to Mimi 1888 1, 0; by Barcaldine ex Lord Lyon mare; inbred to Stockwell at the 3rd & 4th removes.

***Sea Hawk II** 1963 imp ex Fr. '66, a grey stayer by Herbager ex a mare by Free Man; sired Erimo Hawk GC, Matahawk GP and Bruni L Exp. Jap. '63.

†**Shred** 1929 (Lup.), by Teddy ex Pommern mare.

Shut Out 1939 Belmont & Kentucky D; by Equipoise ex Chicle mare.

Silvern 1917, won Coronation Cup & 8 other races; by Polymelus ex Silver Fowl by Wildfowler; inbred to Isonomy at the 4th remove; sired also the dam of Solar Bear.

Sir Archibald 1905, by Desmond ex Arc Light by Prism; inbred to Hermit at the 4th remove.

Sir Walter Raleigh 1928, won 7 unimportant races up to 10f; by Prince Galahad ex Black Jester mare; inbred to St. Simon at the 4th remove.

Skymaster 1958 sprinter by Golden Cloud ex Disclipliner sire of dam of Roland Gardens 1975 2. Exp. Jap. '72.

*†**Solar Slipper** 1945 Ch. & 4 other races; by Windsor Slipper ex Solar Flower by Solario; exp. USA 1956. Sired also Pandofell GC.

***Solonaway** 1946, (Ir. 2) & 5 other good races; by Solferino ex Grand Glacier mare; inbred to Polymelus at the 4th rem.; sired also Sweet Solera 1,0. Exp. Jap. '58.

Sparkler 1968 miler by Hard Tack ex a Preciptic mare, sired Enstone Spark 1975 1.

*†**Spion Kop** 1917 D, by Spearmint ex Gallinule mare; sired also *†Felstead D & the dam of †His Highness.

***Stratford** 1919, miler, won 6 races; by Swynford ex Lesbia by St. Frusquin; inbred to St. Simon at the 3rd & 4th removes.

Supremus 1922, notable USA sire, by Ultimus ex John o'Gaunt mare; inbred to Domino at the 3rd & to Isonomy at the 4th removes.

Tai Yang 1930, stayer, by Solario ex Lemberg mare.

***Telerifique** 1934, French mid. dist., won 5 races; by Bacteriophage ex St. Just mare; sired *†Alizier and the dam of *Le Haar.

Thrush 1902, by Missel Thrush ex Charibert mare.

†**Torpoint** 1900, great stayer, by Trenton ex Sheen mare; sired also the dams of *†Buchan E(twice), Ch., Saltash E and Plantago.

Tremola 1918, miler h'cpr., by Tredennis ex Winkfield mare; inbred to Hermit at the 3rd & 4th removes.

*†**Vandale** 1943, French stayer, won 5 good races; by Plassy ex a mare by La Farina; sired also Douve FO, *Herbager FD & the dam of Phaeton GP.

Vermilion Pencil 1922, stayer, won Chester Vase & Queen Alexandra Stks; by Gainsborough ex Rectify by William The Third; at stud first in France (exp. 1926) then in Argentina; inbred to St. Simon at the 3rd & 4th and again to Galopin at the 4th rem.

Verwood 1910, French, very fast; by Grey Plume ex Ladas mare; inbred to Hampton at the 3rd & 4th removes.

Appendix B

Important horses of the period which have not appeared in the foregoing pages, not being closely related to the mares, their offspring or the lines of stallions.

Surefoot 1887 2, E by Wisdom ex a Galopin mare, sired the d of *†Stedfast Ch.

***†Childwick** 1890 (ces) by St. Simon ex Plaisanterie (ces & cam at 3y) by Wellingtonia; sired La Camargo (see below), Clyde FO, Negofol FD, 3 other useful horses and the dams of 2 further classic wnrs.

Galtee More 1894 2 D L, exp. Russia 1898; by Kendal ex Morganette by Springfield; inbred to Thormanby at the 3rd & 4th removes. Sired Fervor.

Roxelane 1894 FO (F1), by War Dance ex Rose of York (p.87, 188, App. D) by Speculum; inbred to Vedette at the 3rd & 4th removes; dam of Roi Herode (p.127, App. C & D).

Le Samaritain 1895 (FL) a small grey stayer by Le Sancy ex Clementine (exp. Fr. '82) by Doncaster. Sired *†Roi Herode (p.127, App. C & D) and *Isard II.

Musa 1896 O by Martagon ex Palmflower (App. D) by The Palmer; dam of Mirska O. (Exp. Fr. 1909).

Semendria 1897 GP FO (F1), by Le Sancy ex Czardas (dam of Ragotsky FD GP) by Kisber; inbred to Wild Dayrell at the 4th remove; 3rd dam of Biribi (Appx. A)

La Camargo 1898 FO (F1) (Cadran), ran 32 times, won 24; by Childwick ex Jolly Friar mare; 2nd dam of Massine (Appx. A).

Ard Patrick 1899 D, E, Ch. exp. Germany 1903; by St. Florian ex Morganette (d. of Galtee More vs); sired Ariel and the dams of Herald and Alchemist.

Signorinetta 1905 D & O, bred accidentally by Chaleureux when her dam Signorina (by St. Simon) was walking to visit her intended, *Isinglass. In 1918 this remarkable female produced the useful *Winter King, the sire of *Barneveldt GP.

Cherimoya 1908 O (her only start) by Cherry Tree ex Svelte (gd. dam of Sonsie Queen (p.25) (by St. Simon); inbred to Stirling at 3rd & 4th rems; 2nd dam of Cameronian (p.100, App. C & D) and of the Macnab (App. A).

***†Manna** 1922 2 D, broke down in Leger; by Phalaris ex Waffles (dam of Sandwich L) by Buckwheat; inbred to St. Simon at the 3rd & twice at the 4th remove; sired Miracle E, Colombo 2 and the dams of Pasch 2 E, Rising Light & Macherio (It. 2, L).

***†Belfonds** 1922 FD, exp. Argentina by Isard II ex Perth mare; sired Commanderie FO GP, Vendange FO, Peniche FO, Lysistrata FO & the dam of Pont L'Eveque D.

Fairy Legend 1924 FO & her sister **Mary Legend** 1925 FO. by Dark Legend ex Phoenyx mare; inbred to Hampton at the 4th remove; Mary was dam of Feerie FO.

***†Mon Talisman** 1924 FD AT, captured by the Germans and disappeared; by Craig An Eran ex Ruthene (exp. France 1923) by Lemberg; inbred to Cyllene at the 3rd remove; sired Clairvoyant FD GP & the dam of Orfeo GP. Exp. Fr. '23.

Rose of England 1927 O, by Teddy ex Perce Neige by Neil Gow; dam of Chumleigh L.

Boswell 1933 L, E by Bosworth from Flying Gal (exp. USA '40) by Sir Galahad III.

Mistress Ford 1933 FO, by Blandford ex Polly Flinders II (imp. ex. France 1933) by Teddy; dam of Mystery IX E.

Pasch 1935 2, E (3y) by Blandford ex Pasca (exp. Fr. '37) by Manna.

†Le Pacha 1938 FD GP AT (Hoc. Lup. FL), by Biribi ex Advertencia by Ksar; inbred to Chouberski at the 3rd & 4th rems.

†Verso II 1940 FD AT (Hoc. FL), by Pinceau ex Variete by La Farina; sired Lavandin D & the dam of *Le

Fabuleux FD.

My Love 1945 D, GP by Vatellor ex For My Love by Amfortas; inbred Teddy 3 & 4 rem. Exp. Arg. '54.

Nimbus 1946 2, D by Nearco ex Kong by Baytown. Exp. Jap. '62.

Hill Gail (USA) 1949 (Ken. D) by Bull Lea ex Jane Gail by Blenheim. Sired Martial 2 (p. 8). Imp. UK '54.

***Pinza** 1950 D, GVI by **Chanteur** II (Fr) (imp. UK '46) ex Pasqua by Donatello II. Chanteur II also sired Cantelo L, Only For Life 2 (p.62) and the dam of Hurry Harriet Ch. Pinza was inbred Blandford at 4th rem.

***Le Haar** 1954, French 10f horse by Vieux Manoir ex Telerifique mare; inbred to Blandford at the 3rd & 4th remove; sired *Exbury AT & Niksar 2.

*†**Bald Eagle** 1955 won Wash. Int. twice, by Nasrullah ex a mare by Tiger; sired the filly San San AT and the dam of Exceller GP.

Espresso 1958 a tough little stayer who won the Grosser Preis von Baden (12f) twice and several big h'cps. in UK. By Acropolis ex a Bahram mare, he has honoured his parents, benefitted British breeding and distinguished himself by siring the splendid Sagaro (see below) the only horse to win the Gold Cup three years running.

Hula Dancer 1960 1, Ch (3y), trained in France; by Native Dancer ex Flash On by Ambrose Light.

***Faberge II** 1961, miler, by Princely Gift from a mare by Legend Of France, inbred to Blenheim at 3rd & 4th rem. & again to Blandford at 4th, sire of Rheingold AT & Giacometti Ch. Imp. UK '61, exp. Jap. '70.

Meadow Court 1962 IRD, GVI (3y), like some other horses he suffered from being foaled the same year as Sea Bird II. By Court Harwell from a mare by Tom Fool whose dam had won the Argentine D & O. Exp. Can. '69.

Mysterious 1970 1, O by Crepello ex Hill Shade (USA) by Hillary. Hill Shade raced in UK and returned to USA. Exp. USA '75.

Star Appeal 1970 E AT by Appiani II from Sterna (imp. UK from Germany 1966) by Neckar.

Sagaro 1971 GP GC 3 times by Espresso from Zambara by Mossborough, inbred to Bobsleigh at 4th rem. Imp. UK 1977 ex. France.

Grundy 1972 D IRD GVI, by Great Nephew from Word From Lundy by Worden.

Appendix C

IMPORTANT IMPORTS AND EXPORTS SINCE 1880

Date of Move		*from*	*to*	*page*
1882	Clementina	uk	fr	App. B d of Le Samaritain
	The Frisky Matron	uk	fr	101
1883	St. Lucia	uk	fr	179
	St. Angela	uk	fr	148
1884	The Scottish Chief	uk	fr	13 may have ret.
	Tristan	uk	fr	11
1885	St. Blaise 1880 D	uk	usa	
	Bella	uk	fr	d of Heaume '87 FD
1886	Cinderella	uk	usa	177
	Black Corrie	uk	fr	35
1887	The Bard	uk	fr	179
	Traviata	uk	fr	5
1888	War Dance	uk	fr	87
	Your Grace	uk	it	58 ret. '96
1889	Ormonde	uk	arg	148 then '93 to USA
1890	Ossian 1880 L	uk	austria	
1892	Rose of York	uk	fr	107
	Ben my Chree	uk	usa	d of Cap & Bells II O
1893	Princess Iskra	uk	usa	110
	Petrarch	uk	fr	45
	Geheimnis	uk	ger	93
	Harvester 1881 FD	uk	austria	
	Bijou	uk	fr	179
1894	Goldfinch	uk	usa	39
	Reve d'or 1884 1,O	uk	fr	
	Primrose Dame	uk	fr	161
1895	Americus	usa	uk	103
	Galliard	uk	ger	35, 172
	Carbine	nz	uk	43, 49
1896	Trenton	nz	uk	97
	Rhoda B	usa	uk	5
1897	Bona Vista	uk	austria	151
	Yours	it	uk	71
	Sibola	usa	uk	99
1898	Stormy Petrel	uk	fr	95
	Galtee More	uk	rus	App. B
	Kendal Beauty	uk	fr	150
	Disraeli 1895 2	uk	fr	
	Doricles 1898 L	uk	fr	sire of Consols
	Winkfield's Pride	uk	fr	do of Quo Vadis GP
1899	Satiety	uk	ger	23, 164
	Surefoot 1887 2,E	uk	fr	App. B
	Nun Nicer 1895 1	uk	ger	
	Airs and Graces 1895 O	uk	fr	
	Agnes Osborne	uk	Aust-hung	d of Handicapper 2
	†Soliman	uk	fr	
1900	Flying Fox	uk	fr	148
	Rosewood	uk	fr	170
	Hazlehatch	uk	ger	55
	Robert le Diable 1899	fr	uk	sire of Wrack
1901	Kendal	uk	S.am	119
	Dieudonne 1895 Ch	uk	usa	
	Solitaire 1896 Ch	uk	usa	
1902	Merry Token	uk	usa	177
	Janissary	uk	Aust-hung	sire of Jeddah D
	Childwick	uk	fr	App. B
	Sanctimony	uk	fr	101
	Mrs Kendal	usa	uk	d of Atmah 1908 1
1903	Ard Patrick	uk	ger	App. B
	Fairy Gold	uk	usa	12, 177
1903	Isoletta	uk	it	29, 172
	St. Celestra	uk	fr	85
	Ondulee	uk	usa	89, 180
	Hautesse	uk	usa	157
1904	Handicapper 1898 2	uk	bel	
	Cheers 1899 E (3y)	uk	rus	
	Ballantrae	uk	usa	180
1905	Throwaway 1899 GC	uk	fr	
1906	Adam	fr	usa	149 thence to Austria '08
	Rondeau	uk	fr	149
	Rocksand	uk	usa	177 thence to Fr '12
	Orion 1888 Ch	uk	Aust-hung	89
	Diamond Jubilee	uk	arg	161
	Chaleureux	uk	fr	App. B. sire of Signorinetta
	Topiary	uk	usa	93
	Paradoxical	uk	bel	19
1907	Roi Herode	fr	uk	127
1904	Rabelais	uk	fr	24, 164, 166
1905	Bay Ronald	uk	fr	149, 172
1906	Stream of Gold	uk	ger	77
	Irish Lad	usa	fr	89
	Pietermaritzberg	uk	arg	21
1908	Cyllene	uk	arg	9, 151
	St. Lucre	uk	fr	12
	Membrilla	uk	fr	27
	The Admirable Crichton	uk	fr	33
	Quilda	fr	uk	87
	Frizette	usa	fr	89-91, 180
	Merman 1892 GC	uk	ger	
	Wildfowler	uk	fr	107
1907	Navaretta	uk	fr	79
	Gorgos 1903 2	uk	fr	
1909	Musa 1896 O	uk	fr	App. B
	Galvani 1904 Ch (3y)	uk	fr	
	Galeazzo	uk	it	21
1910	Tracery	usa	uk	93 to arg '20, ret '23
	Missel Thrush	uk	arg	sire of Llangwym Ch
	Le Samaritain	uk	arg	App. B
1911	Bridge of Canny	uk	arg	60

Date of Move		from	to	page
	Your Majesty	uk	arg	71
	Sundridge	uk	fr	11 ret after '19
1912	Helicon	uk	fr	137, 166
	Vaya	uk	fr	167
	Ferula	uk	fr	12
1913	Cheery	uk	ger	
	Minoru	uk	rus	81
	Aboyeur 1910 D*	uk	rus	*on disq. of Craganour
	Craganour	uk	arg	32
	Dark Ronald	uk	ger	69
	Cooee	uk	fr	97
	Queen Elizabeth II	uk	fr	173
1914	Lady of Pedigree	uk	fr	
	Omar Khayyam	uk	usa	sired d of Hycilla Ch O
	Polymelian	uk	usa	157
1915	Wrack	uk	usa	40
	Hush	uk	fr	50
	Let Fly	uk	arg	82
	Vaila	uk	usa	178
	Torpenhow	uk	usa	156
1916	Fairy Ray	uk	usa	gd of Gallant Fox ApA
	Light Brigade	uk	usa	60
	Bobolink	uk	usa	41
	Sunstep	uk	usa	40
1917	Dark Legend	uk	fr	12 ret '20 & re-exp fr
	Herodias	uk	usa	94
1919	Prince Chimay	uk	fr	167
	Poet's Star	uk	fr	173
	Troutbeck 1903 L	uk	usa	
	Llangwym 1905 Ch (3y)	uk	fr	
	The Panther 1916 2	uk	arg	
	Doxa	uk	fr	
1918	Night Hawk 1910 L	uk	Australia	36
	Catnip	uk	it	99, 152
1920	Havresac II	fr	it	99, 152, 164
	Chuette	uk	it	164
	Bathing Girl	uk	usa	178
	Brown Prince	uk	usa	sire of Brown Bud ApA
1920	Catherine Swynford	uk	fr	d of Caligula ApA
	Cannobie	uk	fr	App. A
	Dora Agnes	uk	fr	127
	Thomassina	uk	fr	14
	Harry of Hereford	uk	fr	38
	Prince Palatine	uk	usa	89, 157, 161 std fr '17-'20
1919	Pasquita	uk	usa	157
1921	Aleppo	uk	bel	30
	Fair Simone	uk	fr	24
	Marveldt	uk	fr	21
	Kircubbin	uk	fr	10
	Sea Moss	uk	fr	117, 162
	Duccia di Buoninsegnia	uk	it	169
	Clarissimus	uk	fr	169-70
	Comrade	uk	fr	44
	Angelina	uk	it	d of Apelle App. A
	Casquetts	uk	fr	App. A gd of Pharis II p.85
	Titanite	fr	uk	86 ret fr '23
1922	Captivating Stella	uk	fr	17
	Rosedrop	uk	usa	172
	Cylgad	uk	fr	App. A

Date of Move		from	to	page
	Polemarch 1918 L	uk	arg	
	Desna	uk	Australia	150
	Fairy King	uk	ger	App A
	Vermilion Pencil	uk	fr	App. A to arg '31
1923	Mon Talisman	uk	fr	App. B
	Regal Roman	uk	usa	60
	Pearl Maiden	uk	fr	119 after frequent visits
	Jean Gow	uk	usa	166
	Ruthene	uk	fr	App. B d of Mon Talisman
	Spike Island	uk	arg	112 then to fr, then to it.
	Snow Martin	uk	fr	44
	Sourabaya	uk	fr	44
	Black Domino	uk	fr	d of Barneveldt GP
	Stefan The Great	uk	usa	10 ret UK later
	Diadumenos	uk	ger	6
1924	Knockaney Bridge	uk	usa	78
	Volta	uk	usa	105
	Seaweed	uk	fr	117
	Incense	uk	fr	32
	Caligula 1917 L	uk	ger	
	Princess Sterling	uk	usa	d of St. Louis 1919 2
	Saltash	uk	Australia	66
	Pot au Feu 1921 FD	fr	uk	to arg '27
	Call of the Wild	uk	arg	97
	Casteline	uk	fr	d of Casterari
1924	Rose Prince	fr	uk	89, 161
1925	Lady Cicero	uk	fr	60
	Lovelly Clovelly	uk	fr	50
	La Grisette	uk	usa	155
	Roseway 1916 1	uk	usa	
	Queen Silver	uk	usa	d of Silver Urn 1919 1
	Parth	uk	fr	27
	Perfect Peach	uk	usa	10
	Hurry Off	uk	fr	50
1926	Vermilion Pencil	uk	fr	App. A thence to arg '31
	Veilleuse	uk	fr	106
	Muci	fr	it	120
	Apelle	it	uk	131, 165
	Desra	uk	fr	150
	Sir Galahad III	fr	usa	46
	Durbah II	uk	usa	89, 180
	Sunny Jane	uk	usa	66
	Charlebelle 1917 O	uk	usa	
	By George	uk	can	App. A
	Zionist 1922 (ird)	uk	fr	later to bel
	Roseola	uk	fr	d of Tornado p 58
	Grosvenor	uk	Australia	67
1927	Winkipop	uk	usa	4
	Adam's Apple 1924	uk	arg	
	Herbalist	uk	fr	10,
1928	Prince Rose	uk	bel	162 as a foal, to fr'38
	Aileen	fr	uk	143
	Captain Cuttle	uk	it	104
	Pharos	uk	fr	56, 152, 173
	Polly Flinders II	fr	uk	d of Mistress Ford Ap. B
	Ranai	fr	uk	d of Watling Street 1939 D
	Feridoon	uk	fr	150
	Abbots Speed	uk	fr	163
	Argentee	uk	fr	d of Maravedis

Date of Move		*from*	*to*	*page*
	Pharamond II	uk	usa	App. A
	Sweet Lavender	uk	fr	154
	Pharmacie	uk	fr	162
1929	Pride of Hainault	uk	fr	129
	Cinq a sept	uk	fr	98
	Sickle	uk	usa	156
	Franklin 1918 Ch	uk	Austria	
	Quick Change	uk	fr	gd of Steady Aim 1943 O
	Hunter's Moon 1926	uk	arg	By Hurry On ex Selene
	Royal Minstrel 1925 E	uk	usa	
	Sceptical	uk	usa	d of Chop Chop
192–	Princess Palatine	fr	usa	91
–	Belfonds	fr	arg	App. B
1930	Ferry	uk	pol	83
	Lady Maureen	uk	fr	112
	Plucky Liege	uk	fr	46
	Bull Dog	fr	usa	46
	Santorb 1921 GC	uk	hung	
	Priscilla Carter	usa	uk	d of Hycilla 1941, O, Ch.
193–	Strip the Willow 1529 FD GP	fr	arg	
1930	Ensoleillee	uk	fr	20
	Sweet Picture	uk	usa	gd of Talma II 1948 L
193–	Ortello	it	usa	112
193–	Rodosto	fr	arg	App. A
1931	La Boni	uk	fr	52 ret '37
	La Mauri	uk	fr	52
	La Troienne	fr	usa	20, 137
	Teddy	fr	usa	129, 149
	Rustom Pasha	uk	fr	6 later to arg
	The Winter King 1918	uk	fr	ex Signorinetta
1931	Mah Mahal	uk	fr	126
	Nipisiquit 1924	uk	fr	to usa '34
1932	Comedienne	uk	fr	80
	Brulette	fr	uk	118
	Bayora			
	Becti	uk	fr	33
	Tolbooth	uk	it	40
	Blenheim	uk	fr	125, 152, 169
	Tofanella	uk	it	165
	Booklaw	uk	fr	40
1933	Brown Betty	uk	usa	60
1932	Gold Bridge	fr	uk	App A
1933	Mistress Ford	fr	uk	App B
	Captain's Fancy	uk	fr	106
	Artist's Proof	uk	fr	94, 159 to hung '38
	Kopje	uk	fr	74
	Empire Crusade	uk	fr	70
	Foxhunter	uk	fr	119
	Dolores	uk	fr	gd of Souverain 1943 GP, GC
	Najmi	uk	fr	gd of My Love 1945 D
	Imagery	uk	fr	106
1934	Mumtaz begum	fr	uk	126, 152, 153
	Gay Bird	uk	usa	41, 180
	Even Scales	uk	fr	104
	Voleuse	uk	fr	106
	Moirheen Rhue	uk	it	112 & her dau Mannozza
	Firdaussi	uk	fr	App. A
	Rhodes Scholar	fr	uk	40 exp usa '40

Date of Move		*from*	*to*	*page*
	Ponteba	fr	uk	d of Pont L'Eveque D
	Mahmoud	fr	uk	126 to usa '40
	Mary Tudor II	fr	uk	173
	Mirawala	uk	fr	33
	Easton 1931	fr	uk	to usa '40
1935	Piazetta	uk	fr	83
	Columba	uk	fr	41, 180
	Cosquilla	uk	fr	163 ret UK, foaled Princequillo (exp USA) then to usa '40
	Ksar	fr	usa	179
	Diophon	uk	greece	6
	Diolite	uk	jap	App. A
	Friar's Daughter	uk	fr	47, 171
	Diligence	uk	rus	App. A
	Bella Minna	uk	it	165 & again '37
1936	Apulia	uk	fr	70
	Pearlweed	fr	uk	120 to arg '44
	Kantar	fr	usa	133
	Lavendula II	uk	fr	154
	Colour Bar	uk	fr	6
	Sind	uk	fr	33, 166 to arg '40
	Salmon Trout	uk	s.af	44
	Plassy	uk	fr	56
	Hotweed	fr	uk	118
	Santa Cruz	uk	fr	61
	The Recorder 1927	uk	rus	104
	Sundry	uk	fr	d of Coup de Lyon App.A
	Mieuxce	fr	uk	170
1937	Barbara Burrini	uk	it	165
	Buchanite	uk	fr	110
	Sister Anne	uk	fr	74
	Dulce II	fr	uk	128
	Badruddin	fr	arg	126
	Mumtaz Mahal	uk	fr	104, 125, 152 finally
	Cos	uk	fr	6
	Le Ksar 1934 2	fr	uk	to arg '43
	Pasca	uk	fr	d of Pasch App.B
1937	Tete a Tete	uk	fr	98
	Silver Fox II	fr	uk	120
1938	Sabla II	uk	it	52
	Boswell	uk	usa	App. B
	Bois Roussel	fr	uk	46, 174
	Nearco	it	uk	100, 152, 174, 176
	Celerina	uk	fr	86
	Donatello II	fr	uk	74, 169, 170, 173
	Bouillabaisse	uk	fr	33
	Legend of France	fr	uk	
1939	Chor Bazar	uk	fr	106
	Fair Copy	uk	fr	83, ret uk '48, to Swe '56
	Quince	uk	fr	41
	Cillas	fr	uk	91
	Indiscretion	uk	fr	72
	Zulaikha	uk	fr	d of Sandjar 1944 FD
	Atout Maitre	fr	uk	108
193–	Aethelstan	fr	usa	149
1940	Bahram	uk	usa	47, 171
	Heliopolis	uk	usa	61
	Coronach	uk	nz	100
	Crapom	it	ger	App. A

Date of Move		from	to	page
	Avella	fr	uk	46
1941	Cameronian	uk	arg	100
	Hydroplane II	uk	usa	54
	Loaningdale	uk	urag	16
	Sandwich	uk	braz	App. A
	Hycilla 1941 O, Ch	uk	usa	
	Princequillo	uk	usa	163
	Scarlet Tiger	uk	arg	App. A
	Moonlight Run 1938	uk	usa	
1942	Finis 1935 GC	uk	nz	
	Pont L'Eveque 1937 D	uk	arg	
	Bala Hissar	fr	braz	106
1943	King Salmon	uk	braz	141
	Singapore	uk	braz	14, 155
1944	Felicitation	uk	braz	App. A
	Sandjar 1944 FD	fr	braz	
	Namesake	uk	fr	30
1945	Chulmleigh 1934 2	uk	arg	
	High Peak	uk	Australia	16
1946	Nogara	it	uk	100, 158
	Chanteur II 1942	fr	uk	sire of Pinza App. B
1947	Hierocles	uk	usa	App. A
	Prince Chevalier	fr	uk	86, 160, 163
	Gulf Stream	uk	arg	61
	Umiddad	uk	fr	50
	Pay Up	uk	fr	41
	Goya II	fr	usa	123
	Miss France	fr	uk	122
1948	Nicolo dell'Arca	it	uk	100
	Pearl Diver	fr	uk	120 exp Jap '57
1949	El Greco	it	fr	165
	Naucide	it	uk	99
	Red Ray	uk	usa	72
	Nakamuro	it	uk	100 ret it '54
	Delirium 1945	uk	fr	
194-	Source Sucree	fr	uk	154 to usa '53
	Priam II	fr	usa	90
	Macherio	it	usa	112
1950	Nasrullah	uk	usa	126, 152
	Tabriz	uk	fr	52
	Midday Sun	uk	nz	61
	Her Slipper	uk	fr	64
	Nomellini	it	uk	99
1950	Divel	uk	fr	116
	Relic	usa	fr	178 & to UK '57
	Djeddah	uk	usa	90, 153
1951	Tenerani	it	uk	165 ret it '60
	Seven Seas	uk	usa	61
	Satanella	uk	fr	46
	Amboyna	uk	fr	135
	Flocon	fr	uk	120 exp Hol '58
	Pardal	fr	uk	20
1952	Malindi	uk	fr	126
	Knight's Daughter	uk	usa	142
	Turn To	fr	usa	154
	Double Deal II	uk	usa	38
	Never Say Die	usa	uk	40, 126, 152
	Turkhan	uk	fr	174
	Sakountala	fr	uk	180

Date of Move		from	to	page
	Watling Street 1939 D	uk	usa	
	Ujiji 1939 GC	uk	swe	
	Garden Path 1941 2	uk	it	sis to Watling St.
	Rose of Yeroda	uk	fr	94 to usa '56
	Niobe II	usa	uk	d of Aurelius '58 L
	Dynamiter '48 Ch (twice)	uk	fr	
	Merry Boy	fr	uk	83
	Hugh Lupus	fr	uk	180
	Beau Prince II	uk	fr	App. A
	Tyrone 1952 (F2)	fr	uk	
1953	Royal Charger	uk	usa	126, 154
	Lady Angela	uk	can	112, 155
	Sunset III	uk	usa	4
	Souverain '43 GP, GC	fr	uk	
	Shikampur	uk	fr	128
1954	Coaraze	fr	braz	123
	Bell of Troy	usa	uk	138
	Barley Corn	uk	fr	144
1945	Black Tarquin 1945	usa	uk	91 exp usa '50, ret '54
1954	Norseman	fr	usa	106 ret fr '56
	Hill Gail	usa	uk	App. B
	Menetrier 1944	uk	can	
	My Love	fr	arg	App. B
	Masaka 1945 O	uk	usa	
	Majideh	uk	usa	d of Masaka
	Peter Flower 1946 Ch	uk	usa	
	Epigram	uk	den	44
	Souepi 1948 GC	uk	chile	
	Bray Melody	uk	usa	70
1955	Society's Way	uk	peru	98
	Foretaste	uk	fr	142
	Kingsway 1940 2	uk	usa	
	Queen of Basrah	uk	fr	67
	Mah Iran	uk	usa	126
	Monsoon	uk	usa	134
	Our Babu	uk	usa	78 to jap '67
1950	Philippa II	uk	fr	130
1955	Sunset Gun	uk	usa	113
1956	Star of Iran	uk	usa	126
	Eastern Grandeur	uk	usa	33
	Arctic Prince	uk	usa	75
	Ribot	it	uk	165 ret it '57, to usa '60
	Tulyar	uk	usa	16
	Solar Slipper	uk	usa	App A
	Fantan II	usa	uk	165 ret uk '63
	My Babu	uk	usa	78
	Tropique	fr	uk	17 ret fr '63
	Great Captain	usa	uk	138
1957	Omelia	uk	usa	126
	Montaval 1953 GVI	fr	uk	to jap '61
	La Miramboule	fr	usa	d of Tambourine II IRD & Nasram II GVI
	King's Cross	uk	usa	70
	Avenger 1944 GP	uk	usa	
	Rapace 1952 FD	fr	uk	to jap '67
	Pirate King 1953	uk	Australia	sire of Random Shot GC
	Macherio	it	uk	112
1958	Celtic Ash	uk	usa	135 to jap '70

Date of Move		from	to	page
	Niccolina	uk	fr	40
	Court Martial	uk	usa	4
	Arctic Explorer	uk	Australia	75
	Toulouse Lautrec	uk	usa	165 sire of Margueritte Vernaut Ch
	Solonaway	uk	jap	App. A
	Tudor Melody	uk	usa	41 ret uk '60
	Botticelli 1951 GC	it	uk	ret it '62
	Libra	uk	usa	d of the bros Ribocco '64 IRD L & Ribero '65 IRD L
	Polly's Jet 1953	usa	uk	
1957	Right of Way 1957	uk	arg	sire of the d of Crow L
1959	King of the Tudors	uk	usa	78
	Red God 1954	usa	uk	
	Migoli	uk	usa	126
	Galcador	uk	jap	122
	Bastia	fr	uk	173
	Luminary	uk	usa	6
	Matelda	uk	usa	41
	Tudor Minstrel	uk	usa	104
195-	Ambiorix	fr	usa	165
	Le Petit Prince	fr	arg	130
	Good Luck	fr	braz	27
	Scratch II '47 FD L	fr	arg	
	Aquino II	uk	pol	70
	Thunderhead II	fr	s.af	83
	Argur	fr	arg	33
	Elpenor 1950 GC	fr	braz	
1960	Taboun	fr	uk	67
	Duke's Delight	uk	usa	d of Noblesse O
	Sica Boy 1951 AT	fr	uk	
	Shantung	fr	uk	144
1961	Fastnet Rock	uk	swe	134
	Faberge II	fr	uk	App. B
	Javelot 1956 E	fr	uk	
	Court Harwell	uk	arg	75
	Borealis	uk	swe	135
	Mark Ye Well	usa	uk	to swe '64
	Lauso	fr	uk	112
1962	Irish Lass II	uk	fr	144
	Sipsey Bridge	uk	fr	116
	Hill Shade	usa	uk	d of Mysterious 1,0 App.B ret usa '72
	Klairon	fr	uk	64
	Match III	fr	uk	67
	Nimbus (1946)	uk	jap	App. B
1963	Forino II 1959	it	uk	to Jap '69
	Noblesse 1960 O	uk	usa	
	Silly Season	usa	uk	38
	George's Girl	uk	usa	d of Pistol Packer FO
	Emerson	braz	fr	88
	Sybil's Niece	uk	it	112
	Royal Record II	usa	uk	138 to Jap '68
1964	Fighting Edie	uk	fr	112
	Herbager	fr	usa	App. B
1964	Sans le Sou	uk	fr	98
	Mourne	fr	uk	96 ret fr '72
	Zarathustra 1951 GC	uk	jap	
1965	Vimy	fr	jap	22

Date of Move		from	to	page
	White Fire III	fr	uk	116 to jap '69
	Breakspear II 1961	usa	uk	to fr '70
1966	Ambiguity	uk	usa	4
1965	Gilles de Retz 1953 2	uk	jap	
	Carrozza	uk	usa	36
	Pandofell 1957 GC	uk	rus	
	Ratification	uk	jap	22
1966	Hornbeam	uk	swe	4, 181
	Provoke	uk	rus	41
	Star Appeal	ger	uk	App. B
	Aunt Edith 1962 GVI	uk	usa	
	Fair Edith	uk	usa	d of Aunt Edith vs
	Ribero 1965 IRD, L	usa	uk	
	Sir Ivor	usa	uk	60 ret usa '70
	Democratic	uk	arg	App. A
	Sea-Hawk II	fr	uk	App. A
	Reliance II	fr	uk	67
	Indiana 1961 L	uk	jap	
	Kashmir II	uk	fr	104
1967	Larkspur	uk	jap	135
	Immortality	usa	uk	138 to arg '67
1967	Nelcius	fr	uk	30 to jap '71
	Yrrah jnr.	it	uk	52 to can '69
	Only for Life	uk	jap	75
	Appiani II 1963 (it D)	it	uk	exp fr '73 sire of Star Appeal App. B
	Habitat	usa	uk	154
	Hard Ridden 1955 D	uk	jap	
	Martial	uk	arg	10
	Pieces of Eight	uk	usa	36
	Vienna	uk	fr	75, 173 to jap '72
1968	Parthia	uk	jap	80
	Caro	uk	fr	36
	Abermaid 1959 1	uk	usa	
	Homeward Bound 1961 O	uk	usa	
	King's Bench	uk	jap	70
	Fleet	uk	usa	138
	Parbury 1963 GC	uk	Chile	
	Astec 1964 FD	fr	uk	to jap '72
	Ribocco 1964 IRD,L	usa	jap	
	Nijinsky	usa	uk	155 ret usa '71
	Astuce	fr	usa	102
1969	Mill Reef	usa	uk	72, 153
	Primera	uk	jap	108
	Prince Regent	fr	uk	70
	Miralgo 1959 (hard)	uk	jap	sire of Roll of Honour GP
	Lakewoods	uk	usa	22
	Pleben 1969 GP	uk	fr	
	Meadow Court	uk	can	App. B
196-	Lavandin (Fr) 1953 D	fr	jap	
	Monade 1959 O	fr	usa	
	Nasram II 1960 GVI	fr	usa	
	Sea Bird II	fr	usa	6, 158 ret fr later
	Bald Eagle	usa	fr	App. B
1970	St. Crespin III	uk	jap	16
	Intermezzo	uk	jap	57
	Psidium	uk	arg	74
	Espresso	uk	ger	App. B

Date of Move		from	to	page
	Sassafras	fr	uk	144
1970	Hard to Beat	uk	fr	33
1970	Rockavon	uk	fr	6
	Hula Dancer 1960 1, Ch	uk	usa	App. B
1971	Sunny Way	uk	ven	128
	Aggressor 1955 GVI	uk	it	
	Khalkis 1960 E	uk	ger	
	High Hat 1957	uk	jap	sire of Glad rags 1
	Full Dress II	uk	usa	67 ret & exp again usa '76
	Irish Ball	fr	uk	144 to jap '74
	Morston	fr	uk	132
1967	Vienna	uk	fr	75, 174 thence to jap '72
1972	Snob	fr	jap	96
	Skymaster	uk	jap	App. A
	Sodium	uk	fr	4
	Waterloo	uk	usa	22
1973	Sir Gaylord	usa	fr	98, 154

Date of Move		from	to	page
	Roberto	uk	usa	154
	Ginevra 1969 O	uk	jap	
	Scottish Rifle 1969 E	uk	usa	
	Hurry Harriet	uk	usa	52 to can '76
	Darling Boy 1958	uk	hol	sire of d of Mrs McArdy 1
1974	Boucher	usa	uk	138
	Steel Pulse	uk	Australia	102
	Snow Knight	uk	can	82
	Athens Wood 1968 L	uk	rus	
1975	Supreme Sovereign	fr	uk	134
	Giacometti	uk	usa	26
	Roll of Honour 1967 GP	uk	fr	to Turkey '76
	Mysterious	uk	usa	App. B
	Mon Fils	uk	fr	38
1976	Charlottown	uk	Australia	140
1977	Sagaro	fr	uk	App. B
1972	Fair Arabella	usa	uk	d of Fair Salinia 1975 O

Appendix D

IN-BREEDING

In the foregoing pages I have tried to note the incidences of in-breeding up to the 4th generation. There now follow three lists, firstly those important animals closely inbred, secondly those inbred in the 4th generation and thirdly the outbreds (sic) with no in-breeding for 4 generations.

The Close Inbreds

Date		*Inbred to*	*Generations*
1880 to 1899	Melton	Stockwell	3 4
	Gallinule	Stockwell	3 4
	Rinovata	Rataplan	3 4
	Fascination	King Tom	3 4
	We Shall Remember	Stockwell	3 3 4
	Roxelane	Vedette	3 4
	Flying Fox	Galopin	2 3
	Glasalt	Lord Clifden	3 4
		Stockwell	4 4
	Satirical	Stockwell	3 4
	Sirenia	Sis. to Ryshworth	3 3
		Stockwell	4 4 5 5
	Eryholme	Newminster	3 4
		Galopin	3 4
	Marco	'Darling's Dam'	3 4
	Le Sancy	Windhound	3 4
	Sainfoin	Stockwell	3 3
1900 to 1909	Our Lassie	Galopin	3 4
	Valve	Galopin	3 4
	Ajax	Galopin	3 4
	Marcovil	Hermit	3 3
	Bayardo	Galopin	2 4
		Sterling	4 4
	Prince Palatine	Hampton	3 4
	Verve	Hampton	3 4
	Bachelor's D'ble	Hermit	3 4
	Maid of the Mist	Bend Or	3 3
	Brig of Ayr	Galopin	3 3
	Cherimoya	Sterling	3 4
	Electra	Sterling	3 4
	Lemberg	Isonomy	3 3
	White Eagle	Isonomy	2 3
	Scotch Gift	St. Simon	3 3
		Galopin	4 4 4
1910 to 1919	Stefan The Great	St. Simon	3 4
		Hampton	4 4
	Athasi	St. Simon	3 4
	Dolabella	Isonomy	3 4
	Blanche	Isonomy	3 4
		Galopin	3 4
		Sterling	4 4
	Garron Lass	St. Simon	2 3
	Canyon	Galopin	3 4
	Serenissima	Hermit	3 4
	Phalaris	Springfield	3 4
	Havresac II	St. Simon	2 3
	Gainsborough	Galopin	3 4
	Ksar	Omnium II	2 3
	Zariba	St. Simon	3 4
	Polly Flinders	Isonomy	3 4
	Corcyra	Hampton	3 4
		Galopin	4 4
	La Farina	St. Simon	3 4
	Pommern	Hampton	3 3
	Gay Crusader	Galopin	3 3
1920 to 1929	Papyrus	St. Simon	2 3 4
	Manna	St. Simon	3 4 4
	Mon Talisman	Cyllene	3 3
	Gold Bridge	Orby	3 3
	Pharos & Fairway	St. Simon	3 4
	Apelle	St. Simon	3 4
	Pharamond II	Cyllene	3 4
		St. Simon	3 4
	Friar's Daughter	St. Simon	3 4 4
	Uganda	St. Simon	3 4 4
	Colorado & Caerleon	St. Simon	3 4
		Hampton	4 4
	Drift	St. Simon	3 4
	Arabella	Cyllene	3 4
	Duccia di Buon-insegna	St. Simon	3 3
	Fancy Free	St. Simon	3 4 5
	Sickle	St. Simon	3 4
		Cyllene	3 4
	Cavaliere d'Arpino	St. Simon	3 4
		Galopin	4 4
	Vatout	Gallinule	3 4
	Tourbillon	Omnium II	3 4
	Supremus	Domino	3 3
		Isonomy	4 4
	Foxlaw	Donovan	3 4
	Sansovino	Pilgrimage	3 3
	Nogara	St. Simon	3 4
1930 to 1939	Djezima	Rabelais	3 4
	Le Pacha	Chouberski	3 4
	Bois Roussel	St. Simon	3 4
	Carpet Slipper	St. Simon	3 4

Date		Inbred to	Generation
	Big Game	White Eagle	3 4
	Sif	Rabelais	2 3
	Hyperion	St. Simon	3 4
		Galopin	4 4
	Owen Tudor	Chaucer	3 4
	War Relic	Rocksand	3 3
		Fairy Gold	3 3
1940	Mossborough	Chaucer	3 4
to	Clarion	Gay Crusader	3 3
1949	Polynesian	Polymelus	3 4
	Sicambre	Rabelais	3 4
	Alycidon	Swynford	3 4
	Sunny Boy	Gainsborough	3 3
	Tourment	Bruleur	3 4
1950	Le Haar	Blandford	3 4
to	Herbager	Firdaussi	3 3
1959	Tenareze	Colombo	2 3

Date		Inbred to	Generation
	Tenareze (contd.)	Salmon Trout	3 4
	Turn To	Pharos	3 3
	Ballymoss	Gainsborough	3 4
	Dan Cupid	Sickle	2 4
	Hugh Lupus	Tourbillon	2 3
		Bruleur	3 4
	Hethersett	Tourbillon	3 4
	Phil Drake	Teddy	3 4
1960	Vaguely Noble	Hyperion	3 4
+		Bahram	4 4
	Queen's Hussar	Fair Trial	3 3
		Fairway	4 4 4
	Faberge II	Blenheim	3 4
		Blandford	4 4

4th Generation Inbreds

Date		Inbred to	Generation
1880	Sundridge	Stockwell	4 4
to	St. Simon	Voltaire	4 4
1890	Galicia	Voltaire	4 4
	War Dance	The Baron	4 4
	Bona Vista	Pocahontas	4 4
	Cyllene	Stockwell	4 4
	Saraband	Touchstone	4 4
1900	Spearmint	Stockwell	4 4
to	Tracery	Galopin	4 4
1909	Orby	Doncaster	4 4
	Roi Herode	Thormanby	4 4
	Gondolette	Beadsman	4 4
		Madame Eglantine	4 4
	St. Astra	Newminster	4 4
	Veneration II	Thormanby	4 4 4
1910	Jest	Hermit	4 4
to	Tetratema	St. Simon	4 4
1919	Craig An Eran	Bend Or	4 4
	Sunbridge	Galopin	4 4
	Diavolezza	Strathconan	4 4
	The Tetrarch	Doncaster	4 4
		Speculum	4 4
		Rouge Rose	4 4
	Son In Law	Blair Athol	4 4
	Friar Marcus	Hampton	4 4
	Tillywhim	Isonomy	4 4
	Molly Desmond	Hermit	4 4
	Orlass	Galopin	4 4
	Pearl Maiden	Galopin	4 4
		Bend Or	4 4
	Blandford	Isonomy	4 4
	Hurry On	Hermit	4 4
1920	Portlaw	Cyllene	4 4
to		Persimmon	4 4
1929	Singapore	St. Simon	4 4
	Delleana	St. Simon	4 4 4
	Asterus	Hampton	4 4
	Cameronian	St. Simon	4 4

Date		Inbred to	Generation
	Solario	Hampton	4 4
	Pearl Cap	St. Simon	4 4
	Dorina	St. Simon	4 4
	Rose Red	Hermit	4 4 4
	Double Life	Bend Or	4 4
	Aethelstan	Hampton	4 4
	Rialto	Isinglass	4 4
	Blenheim	Isinglass	4 4
1930	Heliopolis	St. Simon	4 4
to	Tideway	St. Simon	4 4
1939	Bozzetto	St. Simon	4 4
	Blue Peter	St. Simon	4 4
	Pharis II	Cyllene	4 4
	Nearco	St. Simon	4 4
	Unbreakable	St. Simon	4 4
		Sainfoin	4 4
	Fair Copy	Cyllene	4 4
	Colombo	St. Simon	4 4
1940	Court Martial	Polymelus	4 4
to	Tulyar	Phalaris	4 4
1949	Straight Deal	St. Frusquin	4 4
	Relic	Rocksand	4 4
		Fairy Gold	4 4
1950	Firestreak	Phalaris	4 4
to	Pall Mall	The Tetrarch	4 4
1959	Klairon	Gay Crusader	4 4
	Le Levanstell	Son In Law	4 4 4
	Hail to Reason	Pharos	4 4
		Plucky Liege	4 4
	Nearctic	Chaucer	4 4
	Meld	Blandford	4 4
	Petite Etoile	Gainsborough	4 4
1960	Ragusa	Pharos	4 4
+	Blakeney	Nearco	4 4
	Brigadier Gerard	Fair Trial	4 4
	Kashmir II	Hyperion	4 4
	Dahlia	Hyperion	4 4

Important horses with no inbreeding for 4 generations

1880 to 1899	Ornament, Fairy Gold, Florizel II, Admiration, Canterbury Pilgrim, Chelandry, St. Frusquin, Concertina, Sceptre, Ormonde, Orme, Persimmon, Isinglass, Bay Ronald, Hastings, La Camargo
1880 to 1899	Kilwarlin, Ladas, Velasquez, Carbine, Martagon, William The Third, Ayrshire, Charm, Dame Masham, Mrs. Butterwick, Stella, Black Cherry, Santa Brigida
1900 to 1909	Donetta, Absurdity, Sunstar, Neil Gow, Anchora, Pretty Polly, Frizette, Sans Souci II, Polymelus, Rabelais, Chaucer, Swynford, Fair Play, Beppo, Chouberski, John o'Gaunt, Alcantara II, Alope, Dark Ronald
1900 to 1909	Charles O'Malley, Popinjay, Popingaol, Plucky Liege, Scapa Flow, Hamoaze, Dutch Mary, Comedienne, Sardanapale, Durban, Banshee, Catnip, Lady Josephine, Lady Juror, Seaweed, Black Toney
1910 to 1919	Teddy, Man o'War, Bruleur, Spion Kop, Diligence, Selene, Buchan, Sun Worship
1920 to 1929	Toboggan, Mirawala, La Troienne, Brulette, Mumtaz Mahal, Blue Larkspur, Brownhylda, Belfonds, Rose of England, Foxhunter, Mah Mahal, Rustom Pasha, Kantar, Deiri, Prince Rose, Sir Galahad III, Bull Dog, Book Law, Dastur, Baby Polly, Coronach, Voleuse, Le Capucin, Hotweed, Bubbles
1930 to 1939	Astronomie, Adargatis, Clarence, Bahram, Overture, Myrobella, Nicolo dell'Arca, Commotion, Alibhai, Sister Sarah, Sun Chariot, Windsor Slipper, Umidwar, Epigram, Ticino, Lost Soul, Mahmoud, Tornado, Phase, Corrida, Goya II, Mumtaz Begum, Rosetta, Feola, Fair Trial, Discovery, Mieuxce, Vatellor, Donatello II, Brantome, Precipitation, War Admiral, Djebel, Badruddin, Rustom Mahal, Aurora, Big Hurry, Bimelech, Fastnet, Admiral Drake, Astrophel
1940 to 1949	Neocracy, Glen Line, Chenille, Prince Chevalier, Tudor Minstrel, Crepuscule, Herringbone
1940 to 1949	Tantieme, Nasrullah, Royal Charger, Petition, Prince Bio, Princequillo, Wild Risk, Worden, Vieux Manoir, Persian Gulf, Vandale, Solar Slipper, Tenerani, Deux Pour Cent, Above Board, Angelola, Abernant, Coaraze, Corejada, Vienna, Honeyway
1950 to 1959	Vali, Relance III, Native Dancer, March Past, Ribot, Crepello, Aureole, St. Paddy, Hasty Road, Bella Paola, Never Say Die, Shantung, Right Royal V, Darius, Sheshoon
1960 +	Sir Ivor, Busted, Northern Dancer, Nijinsky, Sea Bird II, Mill Reef, Grundy, Star Appeal, Royal Palace, Never Bend, Reform, Allez France

Before summarizing these, I should refer to the late Sir Charles Leicester's study in his "Bloodstock Breeding" (1957). While rightly avoiding conclusions, he refers to the following facts:– out of 1000 thoroughbreds *taken at random* 43% were inbred up to and including the 4th generation; he doesn't give dates but they can be assumed to be contemporaneous, they include "winners, non-winners, stallions, broodmares etc." Against these he sets 156 classic winners and 174 leading sires 330 top animals 1900 to 1954) with the following results:–

Inbred *UP TO* the 4th gen. 26%, inbred *AT* the 4th gen. 27%: a total of 53%.

Now compare the 328 important horses in my Ap. D. above (which must include many of Leicester's, though I omit a number of unimportant ones) and they show 32% inbred *UP TO* the 4th gen. and 21% *AT* the 4th: again a total of 53%.

Compare also the late Col. Vuillier's* study of 654 top racers from 1748-1901, on the whole more ancient, which produce a total inbreeding of 54%. Much the same.

Descending now from these Olympic heights to a study of 753 prominant winners from The Statistical Abstract *plus* 485 stallions from Templar's "Sires of Winners", a total of 1238 horses *EXCLUDING ANY OF MY Ap. D animals:* these produce a total of 36% inbreds.

This indicates that there were more inbreds among the very top class horses than among the 'second bests'.

It is possible to produce statistics to support the fact that, during the period 1910 to 1939 when St. Simon's offspring featured close up in the pedigrees of so many horses, more of them were inbred to him than to any other stallion before or after (probably since Stockwell 1849). The foregoing lists surely indicate this and would undoubtedly do so more extensively, had not St. Simon's son, Persimmon, died at the early age of 15.

**Col. Vuillier, the inventor of the 'dosage' system, was adviser on breeding to the late Aga Khan.*

I am dodging the issue of artificial insemination and the colossal impact it must have, however forcefully it is controlled; this book is an attempt at history; prophecy is beyond me. I wonder though, when a highly sophisticated data-bank can supply breeders with "ingredients", data not only of stallions but also of their own mares, whether inbreeding to the great stars will occur to a greater or smaller extent at top level; it surely won't be ruled out.

Epilogue

When, many years ago, I started examining pedigrees, it was rather like stamp collecting, the more the better. At first it was names, names with little meaning; then gradually the names became characters and that's when the fun started.

I've enjoyed compiling these pages and, in bidding farewell I want to say how grateful I am to the Thoroughbred Breeders' Association's wonderful library and those knowledgeable and helpful people who have run and are running it now.

The help I have had from my wife in patiently indexing, proof-reading and correcting many of my mistakes has proved invaluable.

Index